# Solar System Planets and Exoplanets

*Edited by Joseph Bevelacqua*

Published in London, United Kingdom

IntechOpen

*Supporting open minds since 2005*

Solar System Planets and Exoplanets
http://dx.doi.org/10.5772/intechopen.90977
Edited by Joseph Bevelacqua

Contributors
Emilio Ramírez Juidías, Katherine Villavicencio Valero, Arthur Borja, Vincenzo Rizzo, Richard Armstrong, Hong Hua, Nicola Cantasano, Tommaso Niccolò, Giorgio Bianciardi, Seiki Augustine Chiba, Mikio Waki, Bryan Palaszewski, Katrin Krohn, Valentina Zharkova, Hector Pérez-de-Tejada, R. Lundin, Cesare Guaita, Joseph John Bevelacqua



First published in London, United Kingdom, 2021 by IntechOpen
IntechOpen is the global imprint of INTECHOPEN LIMITED, registered in England and Wales, registration number: 11086078, 5 Princes Gate Court, London, SW7 2QJ, United Kingdom
Printed in Croatia

British Library Cataloguing-in-Publication Data
A catalogue record for this book is available from the British Library

Additional hard and PDF copies can be obtained from orders@intechopen.com

Solar System Planets and Exoplanets
Edited by Joseph Bevelacqua
p. cm.
Print ISBN 978-1-83969-312-0
Online ISBN 978-1-83969-313-7
eBook (PDF) ISBN 978-1-83969-314-4

For EU product safety concerns:
IN TECH d.o.o., Prolaz Marije Krucifikse Kozulić 3, 51000 Rijeka, Croatia, info@intechopen.com or visit our website at intechopen.com.

# Meet the editor

Joseph John Bevelacqua, Ph.D., CHP, RRPT, is the President of Bevelacqua Resources, Richland, Washington, USA, a provider of physics-related consulting services. A theoretical nuclear physicist by training, Dr. Bevelacqua is a Certified Health Physicist, Registered Radiation Protection Technologist, and Certified Senior Reactor Operator and has over 45 years of professional experience. He was a key player in the Three Mile Island and Hanford cleanup activities, and he is an active researcher with over 185 publications and 4 textbooks. His research areas include theoretical nuclear physics, theoretical particle physics, cosmic radiation, astrophysics, planetary studies, earth science, solar physics, cancer therapy using heavy ions and microspheres, gravitation, group theory, mathematical physics, and applied health physics. Additional effort has focused on muon and tau colliders and theoretical studies of a tetraquark, pentaquark, and hexaquark systems. Studies involving quark fusion have also been published. Recent research has involved space medicine, physiological and radiological effects of manned space missions, COVID-19 treatment using low-dose radiation therapy, neutron excess nuclei formation in stellar collisions, and superheavy nuclei. He received a California University's Professional Excellence Award for his accomplishments. Dr. Bevelacqua is a member of numerous professional organizations including the American Physical Society, American Nuclear Society, American Mathematical Association, Health Physics Society, and Royal Astronomical Society of Canada.

# Contents

# Preface

*Solar System Planets and Exoplanets* describes planetary systems and their characteristics. Although the text focuses on the solar system, exoplanetary systems are also addressed. The book provides a current view of active planetary research areas with a focus on Mars and recent data and analysis from a variety of space probes.

Chapter 1 outlines the characteristics of solar system planets as well as their moons. The Asteroid Belt, Kuiper Belt, dwarf planets, comets, and meteoroids are also reviewed. An overview of exoplanets, their characteristics, and detection methods are summarized. The results of current and planned space probes, and their data-gathering capabilities are provided.

Chapter 2 describes historical, current, and future solar output characteristics. In particular, millennial oscillations of solar irradiance and magnetic field characteristics spanning the years 600–2600 are provided. This chapter discusses the consequences of these oscillations on the terrestrial atmosphere.

Chapter 3 focuses on Venus. Solar cycle variations in the position of vortex structures in the Venus wake are discussed. This chapter notes the interesting possibility that similar conditions should also be applicable to Mars, other bodies within the solar system, and exoplanets in external stellar systems.

Recent Mars data and associated observations are addressed in Chapters 4–8. Of particular interest are data that suggest the possibility that life could have existed on the planet's surface. The search for life on Mars is one of the main objectives of space missions and is the focus of Chapters 4–6. Technological advances in terms of operating aircraft on Mars (Chapter 7) and propulsion systems (Chapter 8) provide useful commentary regarding future innovations that will enhance upcoming space missions.

Chapter 4 covers sedimentation and algorithms to detect the possible existence of vegetation and humidity in the landing area of the Mars Exploration Rover-B (Opportunity). This chapter suggests the possibility of three types of water on the surface, as well as concentrations of neoxanthin, also on the landing area surface, that suggest the possible existence of microalgae.

Chapter 5 addresses the possibility of the existence of life on Mars. The chapter's morphological and morphometrical investigations suggest the presence of remnants of complex algal-like biota that could have lived on Mars.

Chapter 6 outlines new insights into the search for life on Mars. Analyses of data from a number of space probes suggest the possibility that life could have existed in the distant past of Mars, when its atmosphere was wetter and denser.

The challenge of controlling a small aircraft on the surface of Mars is the topic of Chapter 7. This chapter discusses the control of an aircraft exploring the surface of Mars using dielectric elastomers. A structural model is developed for a wing having a control surface, and a wind tunnel test is completed.

Chapter 8 addresses the utilization of chemical and nuclear electric propulsion for the exploration of the Martian moons. The moons, Phobos and Deimos, have potential resources for refueling future space vehicles. The use of in situ resources has the potential to significantly reduce the dependence on Earth's resources on space vehicle propellants, liquid water, and breathing gases. These local resources offer the potential for space missions beyond Mars, and they offer a significant option for deep space exploration.

Chapter 9 covers unique features of asteroid belt bodies Vesta and Ceres. Special crater types, as revealed by the space probe Dawn, are the topic of this chapter. This chapter presents the three most characteristic crater forms found on both bodies.

The author acknowledges the assistance of Jelena Germuth and Anja Filipović for their assistance during the early stages of this book's development. Sara Debeuc is especially acknowledged for her kind assistance throughout the development, review, and finalization of this work. Her efforts significantly contributed to the finalization of this book.

**Joseph Bevelacqua**
Bevelacqua Resources,
Richland, WA, USA

Section 1

# Introduction

Chapter 1

# Solar System Planets and Exoplanets

*Joseph Bevelacqua*


## Abstract

Solar System planets have been studied for centuries, but the observation of exoplanets is only a few decades old. Consequently, knowledge of exoplanets is considerably more limited than Solar System planets. This chapter reviews the essential characteristics of Solar System planets and associated data derived from a variety of observational approaches. Exoplanet characteristics and their comparison to Solar System planets are provided as well as general detection methods and planned probes to gather additional data.


**Keywords:** Solar System planets, Exoplanets, Dwarf Planets, Moons, Asteroid Belt, and Kuiper Belt

## 1. Introduction

Following the birth of the Universe through the Big Bang, a cyclic creation, or another unique event, space was filled with nebulae composed primarily of gas and dust. Stars formed from this primordial material, and the residual mass or interstellar medium (ISM) formed the constituents that led to planet formation. There are numerous papers, references, and books describing the characteristics of Solar System planets as well as exoplanets [1–35].

This crude model for planetary formation is based on the assumption that a star forms from the gravitational attraction and associated collapse of the primordial material. The contraction of the star with its decreasing radius increased the angular momentum of the accretion disk of ISM that formed around the star [3, 7, 10, 15]. The temperature of the material within the accretion disk varied with distance from the star. This temperature dependence caused rocky bodies to form throughout the disk, but icy bodies developed at greater distances. In the Solar System, the icy bodies developed beyond the Asteroid Belt.

Within the Solar System, the terrestrial planets formed from rocky bodies (i.e., preplanetary clusters also known as planetesimals). The terrestrial planets include Mercury, Venus, Earth, and Mars. The larger planets (i.e., Jupiter, Saturn, Uranus, and Neptune) formed from the rocky bodies, icy bodies, gas, and dust that led to their increased size. The higher temperatures and lower masses of the terrestrial planets limited their capture of gases. This was not the case for the giant planets. For Jupiter and Saturn, the larger masses and cooler temperatures led to the capture of significant atmospheres.

The initial planet structures also developed their own accretion disks that led to the formation of planetary moons. These disks were larger for the giant planets, which led to these bodies generally having more moons than the less massive terrestrial planets. Some moons were formed by planetary gravitational capture of rocky structures and asteroid fragments. Other moons (e.g., Earth's moon) formed when a large body collided with the planet.

Following the creation of the initial planets and their moons, the Solar System still contained considerable debris that collided with these bodies. The Moon's craters are an example of the effect of the resulting impact of this debris. Some of this debris, particularly icy structures, formed beyond Neptune's orbit and as is known as the Kuiper Belt. An Asteroid Belt comprised of rocky structures formed between the orbits of Mars and Jupiter.

This simplified model of planetary creation has been supplemented with a bifurcation model. Within the bifurcation model, planet formation occurred in spatially and temporarily distinct domains through a postulated mechanism that was driven by the presence of water [24]. Although the details of this mechanism are unknown, the domains evolved in distinct physical modes with different volatile materials. Model calculations suggest that these physical differences led to the formation of the terrestrial and gas giant planets.

As the capability to observe exoplanetary systems and their atmospheres improves, it will be interesting to determine if the characteristics of the Solar System and life on Earth are unique. Will further exoplanet observations reveal a variety of star and planetary systems having the capability of sustaining life?

The reader should note that the literature provides a range of values for planetary data including their associated composition. Given this consideration, specific references are cited when particular data are noted. Significant figures are usually provided to accommodate the variation in literature values.

This chapter provides a general overview of Solar System planets and exoplanets. It's intended to introduce these systems to readers not well versed in planetary science. Additional planetary details are provided in this chapter's references.

## 2. Planetary overview

Although Solar System planets have been studied for centuries, the observation of exoplanets is only a few decades old. The first exoplanet orbiting a star outside the Solar System (i.e., 51 Pegasi), discovered by Mayor and Queloz [2], did not occur until 1995. Since that time, progress in exoplanet discovery and characterization has increased rapidly. As of January 2021, 4341 exoplanets have been discovered.

General characteristics of Solar System planets are addressed in Section 2.1 and other Solar System bodies (i.e., planetary moons, the Asteroid Belt, comets, Kuiper Belt Objects, and meteoroids) are discussed in Section 2.2. Section 2.3 is devoted to a discussion of exoplanets. Exoplanet detection methods are provided in Section 3. Section 4 summarizes the variety of space probes utilized to examine planetary and exoplanetary systems.

The presentation of this chapter is designed to facilitate the flow of subsequent material. Selected content provides an overview of subsequent chapters. Since there are numerous references to the characteristics of Solar System planets and exoplanets, the presentation of this chapter is necessarily incomplete. However, the text does provide an overview of the subject and the basis for more detailed study utilizing the accompanying reference list.

### 2.1 Solar System planets

There are eight Solar System planets as well as dwarf planets, planetary moons, the Asteroid Belt, comets, Kuiper Belt Objects, and meteorites. These spatial bodies are addressed in subsequent discussion.

An overview of the Solar System planets and a general classification scheme is summarized in **Table 1**, and illustrated in **Figure 1**. The Solar System planets can be grouped in terms of structural characteristics with the terrestrial planets being compact objects that are primarily rocky objects. Gas giant planets are a combination of gas, rock, and ice.

The terrestrial planets are closest to the Sun and are smaller, warmer, and less massive than the gas giants. Although more massive, the gas giants have a lower density. The gas giants have more moons and exhibit ring structures. These general characteristics are summarized in **Table 1**.

**Table 2** provides additional data for the Solar System planets including their orbital characteristics (i.e., distance from the Sun, orbital period, and orbital eccentricity). Their surface gravitational acceleration in $m/s^2$ is also provided. These data are important for spacecraft and probes attempting to investigate these worlds. Detailed orbital calculations are required to plan missions that would reach these planets, and successfully orbit and land on these worlds [4].

#### *2.1.1 Mercury*

Mercury is the planet closest to the Sun, the smallest of the Solar System planets, and similar in size to Earth's moon. It has no moons, and has a cratered surface that is similar to the topography of Earth's moon. Given its proximity to the Sun and high temperature, Mercury's atmosphere is thin with a constituent number density $\leq 10^{11}$ particles/$m^3$ [15]. Mercury's limited atmosphere consists of oxygen, sodium, potassium, and calcium that evolved from surface material. Thermal effects that increase the temperature of the crust and the impact of solar particles and meteorites on the surface create this atmospheric composition. Limited atmospheric hydrogen and helium are derived from particles emitted by the Sun, and

| Planetary Characteristic | Terrestrial Planet | Giant Planet |
|---|---|---|
| Structural Form | Primarily rock | Combination of gas, rock, and ice |
| Mean distance from Sol (AU) | 0.39–1.5 | 5.2–30 |
| Mean surface temperature (K) | 220–730 | 70–170 |
| Mass ( $M_\oplus$ ) | 0.055–1.0 | 15–320 |
| Equatorial radius ( $R_\oplus$ ) | 0.38–1.0 | 3.9–11 |
| Mean density ($g/cm^3$) | 3.9–5.5 | 0.69–1.6 |
| Period of sidereal rotation at equator | 24 h – 243 d | 9.9 h – 17 h |
| Number of moons[b] | 0–2 | 14–82 |
| Ring systems | No | Yes |

[a] *Adapted from Refs. [1, 15]. Rounded to two significant figures due to various values quoted in the literature.*
[b] *Ref. [28].*

**Table 1.**
*General Solar System Planet Characteristics.*[a,b]

**Figure 1.**
*Solar System planets and the dwarf planet Pluto [35].*

| Planet (Number of moons) | Mass ($M_\oplus$) | Mean distance from Sol (AU) | Orbital eccentricity | Orbital period (y) | Equatorial surface gravitational acceleration (m/s$^2$) |
|---|---|---|---|---|---|
| Mercury (0) | 0.055 | 0.39 | 0.21 | 0.24 | 3.7 |
| Venus (0) | 0.82 | 0.72 | 0.007 | 0.62 | 8.9 |
| Earth (1) | 1.00 | 1.0 | 0.017 | 1.0 | 9.8 |
| Mars (2) | 0.11 | 1.5 | 0.093 | 1.9 | 3.7 |
| Jupiter (79) | 320 | 5.2 | 0.049 | 12 | 23 |
| Saturn (82) | 95 | 9.5 | 0.053 | 29 | 9.1 |
| Uranus (27) | 15 | 19 | 0.046 | 84 | 8.3 |
| Neptune (14) | 17 | 30 | 0.012 | 160 | 11 |

[a]*Adapted from Refs. [1, 3, 4, 10, 15].*
[b]*The number of moons is derived from Ref. [28].*

**Table 2.**
*Solar System Planetary Data.*[a,b]

trapped by Mercury's weak magnetic field. The existence of this field suggests that Mercury has an iron core, and that a portion of it is a rotating liquid [22].

Additional Mercury data has been developed by a number of Solar System probes [11] including Mariner 10 and Messenger. The Japanese probe BepiColombo will provide additional data.

### 2.1.2 Venus

The second planet from the Sun is Venus, and it is somewhat smaller than Earth. In spite of the fact that Venus is the closest planet to the Earth, it is one of the least explored planets in the Solar System. Venus is similar in size and mass to the Earth, but these two planets are very different in their environment including temperatures, surface features, and atmospheric composition. Its thick atmosphere traps heat making Venus the hottest planet in the Solar System. Venus has an atmosphere composed of about 97% carbon dioxide with about 3% nitrogen. Trace amounts of water, carbon monoxide, argon, and sulfur dioxide comprise most of the remaining composition. The surface atmospheric pressure is about 90 times that of Earth [15]. Venus also appears to be have been affected by significant surface volcanic activity. Mariner 2 measured a surface temperature greater than 750 K [7, 15]. Sulfuric acid clouds are predicted by limited probe data [7]. Venus also differs from Earth with a slow retrograde rotation rate of about 240 days and lack of a magnetic field that that is an essential element in shielding a planet from the effects of the Solar wind. In addition, Venus has no moon, and its magnetic field is significantly weaker than Earth's [22].

The possible detection of phosphine in the Venusian atmosphere created interest in further exploration of this planet [21]. Phosphine is an uncommon molecular species that is associated with anerobic bacteria. Its detection raised the possibility of extraterrestrial life. However, further measurements are required to confirm these preliminary observations.

Venus appears to be bright because its heavy cloud cover reflects most of the incident sunlight. Astronomers have been able to cut through the clouds of Venus with radar, because radio waves pass through the clouds and bounce off the surface. Using radar, the NASA probe Magellan assembled a detailed surface map while in orbit around Venus [7]. These radar images suggest that Venus has continents with varied topography including canyons, meteorite craters, and volcanic mountains [11].

### 2.1.3 Earth

Excluding the possible phosphine detection on Venus, Earth is the only Solar System planet that has sustained life. In addition, Earth is the only planet with liquid water residing on its surface, and is the largest of the terrestrial planets. Its surface is composed of about 29% land and 71% water that include oceans, lakes, and rivers [7, 15]. Ice covers much of Earth's Polar Regions.

Earth is protected from charged particle ionizing radiation emitted by Solar flares and other disturbances [4–6, 14, 15] by its geomagnetic field and atmosphere. Charged particles from Solar Particle Events and Galactic Cosmic Radiation are reduced in intensity by the attenuating properties of the atmosphere and the action of the geomagnetic field. The atmosphere also shields Earth from a portion of the Sun's ultraviolet radiation that mitigates its harmful effects. Earth's atmosphere is composed of 78% nitrogen, 21% oxygen, and about 1% argon. Trace gases accounting for about 0.1% of the atmospheric composition includes carbon dioxide, methane, nitrous oxides, and ozone [7, 15].

Earth rotates in a prograde manner with a period of 24 hours. It has a nearly circular orbit with a mean distance from the Sun of 1 AU. Given the current Solar luminosity, this distance creates a positive environment for sustaining liquid water that is an essential element for life to be created and sustained.

The Earth has varied topography that is not static and still experiences periodic volcanic eruptions. Earthquakes and violent weather continue to reshape the Earth's surface.

### 2.1.4 Mars

Mars is about 50% further from the Sun than Earth and about half the Earth's size. The planet Mars is only about 10% as massive as the Earth, and has two small moons [7, 10, 11, 15, 28]. Its distinctive reddish color is attributed to the presence of iron oxide in the crust and surface dust.

Mars' atmosphere is composed primarily of carbon dioxide (95%) and nitrogen (3%) with the remainder consisting of argon with trace amounts of carbon monoxide, methane, oxygen, water, and other gases. Mars is also subjected to dust storms that can be massive as well as severe. Its atmosphere is thin with a surface pressure that is less than 1% of sea level pressure on Earth.

Various surface temperatures are quoted in the literature but lie roughly in the range of 130–290 K [15]. The Polar Regions are considerably colder with ice and solid carbon dioxide forming the polar caps [7, 15].

Mars has a 24.5 hour prograde rotation period that is somewhat larger than Earth's. The surface includes cratered areas, extinct volcanoes, chasms, and areas suggesting the previous existence of the action of water flow. In its past Mars was probably warmer than it is today with significantly more water and a thicker atmosphere.

There are anomalies in the Martian magnetic field [22]. Most of the Northern hemisphere is unmagnitized, but portions of the Southern regions are magnetized. These anomalies suggest that at one time Mars had a magnetic field that was similar to Earth's [22].

A number of probes including NASA's Viking 1 and 2 orbiters, as well as the Hubble Space Telescope provided data regarding the characteristics of Mars. Rovers including NASA's Spirit, Opportunity, and Curiosity added additional surface data [11]. Other probes are likely to begin providing data in 2021. These include NASA's Perseverance, China's Tianwen-1, and the United Arab Emirates' Hope.

The possibility of space missions and colonization of Mars are areas of active research [18]. Considerable research is required to develop spacecraft and Mars habitats to permit these activities to occur in a safe manner. A successful colonization effort must provide food, water, power, and shelter. The space radiation environment including Solar Particle Events and Galactic Cosmic Radiation should also be addressed to ensure the health and safety of the colonists [4, 6, 14, 18].

### *2.1.5 Jupiter*

Jupiter is the largest of the outer four planets. Jupiter and Saturn are often designated gas giants while Uranus and Neptune are ice giants.

**Table 3** provides a comparison of the atmospheres of the Sun and the gas giant planets derived from Refs. [10, 15]. Jupiter and Saturn have average compositions similar to the Sun. Uranus and Neptune are also similar in composition to the Sun, but have a higher fraction of the heavier elements.

| Gas | Atmospheric Elemental Composition (Fractional number density of particles)[a] | | | | |
|---|---|---|---|---|---|
| | Sun | Jupiter | Saturn | Uranus | Neptune |
| Hydrogen | H:0.84 | 0.86 | 0.96 | 0.85 | 0.85 |
| Helium | He:0.20 | 0.16 | 0.034 | 0.18 | 0.18 |
| Water | O:0.0017 | 0.0026 | >0.0017 | >0.0017 | >0.0017 |
| Methane | C:0.00079 | 0.0021 | 0.0045 | 0.024 | 0.035 |
| Ammonia | N:0.00022 | 0.00026 | 0.0005 | <0.00022 | <0.00022 |
| Hydrogen Sulfide | S:0.000037 | ~0.00022 | 0.0004 | 0.00037 | 0.001 |

[a]*Derived from Refs. [10, 15].*

**Table 3.**
*Composition of Giant Planet Atmospheres.*[a]

Observational data suggest that Jupiter and Saturn possibly have rocky cores of magnesium, silicon, and iron with additional icy mass. Numerical simulations, based on limited data [15], suggest that the cores of Jupiter and Saturn have masses of about 10 $M_{\oplus}$ and 15 $M_{\oplus}$, respectively. These values have an uncertainty of about 50% [15].

Jupiter is more than twice as massive as the combined mass of the other Solar System planets. However, its mass is about 0.001 of the Sun's mass. Jupiter is the most rapidly rotating Solar System planet with a rotational period of about 10 hours [11] and has an orbital period of about 12 years.

Jupiter is the largest Solar System planet, and has a number of distinctive features. Its Giant Red Spot is a storm region that is about twice as large as the Earth [15]. The spot has a rotation period of about 10 hours [15].

Jupiter also has very thin rings. The first of these, discovered by Voyager 1, are considerably less prominent than Saturn's [7]. The rings appear to be primarily composed of dust [15].

Galileo discovered its largest moons (Ganymede, Callisto, Europa, and Io) that are discussed in Section 2.2.1. Jupiter has 79 moons [28].

Jupiter's magnetic field is significantly stronger [15, 31] than Earth's field and has a large physical extent. It is likely generated by electrical currents in the planet's core. Jupiter's field extends about $5 \times 10^6$ km in front of the planet and stretches about $650 \times 10^6$ km behind the planet extending beyond the orbit of Saturn [7]. In addition, there are regions of enhanced magnetic flux near the poles that create a complex field structure [22].

The Voyager 1 probe provided an initial view of Jupiter and observed volcanic activity [11]. Pioneer 10 and 11, Voyager 2, NASA's Cassini Space Probe, and the Juno Space Probe have added additional data [11, 15].

#### *2.1.6 Saturn*

Saturn lies beyond Jupiter and is the sixth planet from the Sun. It has a rotational period of about 11 hours that partially explains why Saturn is an oblate spheroid with an equatorial bulge [7]. Saturn's mass is about 95 times Earth's.

Saturn has cloud bands that are similar to those on Jupiter, but they are less colorful. Voyager 1 and 2, the Hubble Space Telescope, and the Cassini Space Mission provided views of the color variations as well as Saturn's ring structures.

**Table 3** lists the elemental compositions of the atmosphere of Saturn. The atmospheric molecular gases include hydrogen, helium, methane, and ammonia.

Saturn has a complex ring system and the largest number of moons (82) of any planet in the Solar System [28]. The Hubble Space Telescope imaged Saturn and provided a unique view of its rings. NASA's Cassini Space Probe provided a detailed view of the rings that indicated their structure was comprised of numerous narrow ringlets [11]. Voyager 1 and 2 obtained additional data. The rings are generally broad and flat, and are composed of ice chunks and rocky material [7].

Saturn's magnetic field is almost symmetric about its axis [22]. It is weaker and less complex than Jupiter's field.

#### *2.1.7 Uranus*

Uranus is the Solar System planet with the third largest diameter [15]. It has an atmosphere dominated by hydrogen, helium, and methane. Ammonia and trace amounts of water and hydrogen sulfide are also present [7, 10, 15]. Although Voyager 2 and the Hubble Space Telescope observed some cloud bands, these are less distinct and colorful than those on Jupiter and Saturn. Uranus' interior generates considerably less thermal energy than Jupiter and Saturn [7].

Uranus has a distinctive orbit that differs from other Solar System planets. Its orbit is unique since the obliquity of Uranus' rotation axis is about 98 degrees. Contrary to the other Solar System planets, Uranus rotates on its side in the orbital plane [7]. The net effect of Uranus' orbital characteristics is that during its 84 year period one pole at a time receives more solar radiation than the equatorial region [7, 15]. However, measurements suggest that Uranus' equatorial region is warmer than its poles [7].

Uranus has a system of multiple rings as well as 27 moons [28]. Voyager 2 and the Hubble Space Telescope observed these ring structures. Uranus' rings are faint and much less dramatic than those on Saturn. The rings also wobble due to an unsymmetrical gravitational field since Uranus has a slightly flattened geometric structure [7].

Uranus also possesses a non-symmetric magnetic field. This field is significantly lopsided [7]. There is currently no viable explanation for this field structure.

#### *2.1.8 Neptune*

Neptune was predicted to exist following a celestial mechanics analysis of anomalies in the expected orbit of Uranus. In 1846, it was discovered near the predicted position. Neptune resides at about 30 AU from the Sun, and its orbital period is about 160 years [7, 15].

As noted in **Table 3**, Neptune has an atmosphere dominated by hydrogen and helium with a smaller amount of methane, and is similar to the atmosphere on Uranus. Voyager 2 revealed a blue color and the fastest winds in the Solar System with speeds reaching 2100 km/h [7]. The blue color is derived from small amounts of methane in its hydrogen and helium atmosphere [7].

As noted in **Table 2**, Neptune has a mass about 17 times the Earth's, and it is about four times wider than Earth [15]. In a manner similar to Jupiter and Saturn, Neptune generates more energy than it receives from the Sun [15].

In a similar manner to the other giant planets, Neptune has a system of multiple rings as well as 14 moons [28]. However, Neptune's rings are not as extensive as Saturn's. Voyager 2 observed that Neptune's rings appear to be composed of a high concentration of dust particles [7].

### 2.2 Additional Solar System Bodies

In addition to planets, there are additional bodies that reside in the Solar System and have the potential to affect planetary motion as well as their environment. These bodies include planetary moons, the Asteroid Belt, comets, Kuiper Belt Objects, and meteorites. Each of these Solar System bodies is addressed in subsequent discussion.

#### *2.2.1 Planetary Moons*

As noted in **Table 2**, the number of moons varies with the Solar System planet and their characteristics [1, 3, 4, 10, 15]. The larger planets tend to have more moons than the smaller ones. However, dwarf planet Pluto has five moons, which is an exception to the aforementioned trend [28, 29]. Its moon Charon is larger than any other Solar System moon relative to its planet size. **Table 4** summarizes the four largest Jupiter moons (Callisto, Io, Europa, and Ganymede), Saturn's largest moon (Titan), Neptune's largest moon (Triton), and Earth's moon that are more massive and physically larger than Charon. Characteristics of these planetary satellites including their mass, radius, density, orbital period, and semimajor axis are summarized in **Table 4**.

| Moon (Planet) | Mass ($10^{22}$ kg) | Radius ($10^3$ km) | Density (g/cm$^3$) | Orbital Period (d) | Semimajor Axis ($10^3$ km) |
|---|---|---|---|---|---|
| Moon (Earth) | 7.3 | 1.7 | 3.4 | 27 | 380 |
| Io (Jupiter) | 8.9 | 1.8 | 3.5 | 1.8 | 420 |
| Europa (Jupiter) | 4.8 | 1.6 | 3.0 | 3.6 | 670 |
| Ganymede (Jupiter) | 15 | 2.6 | 1.9 | 7.2 | 1100 |
| Callisto (Jupiter) | 11 | 2.4 | 1.8 | 17 | 1900 |
| Titan (Saturn) | 13 | 2.6 | 1.9 | 16 | 1200 |
| Triton (Neptune) | 2.1 | 1.4 | 2.1 | 5.9 | 350 |

[a]*Derived from Ref. [15]. Two significant digits are used due to various literature values.*

**Table 4.**
*Data for Selected Solar System Moons.*[a]

### 2.2.1.1 Ganymede

Jupiter's moon Ganymede has a diameter of about 5200 km and is the largest moon in the Solar System. It is more massive than Pluto. Ganymede has an icy surface with numerous craters [7, 15].

Ref. [7] suggests that Ganymede has a structure comprised of three basic layers. The first is an iron core with the possible presence of sulfur. A layer of rock surrounds the core. The final layer is an icy crust that overlays the rocky layer.

Ganymede has an electromagnetic field that is likely generated by currents in its core. The field also has a contribution from inductive effects caused by Jupiter's magnetic field. These effects could be attributed to a layer of salt water residing under the surface ice [7].

The atmosphere of Ganymede is limited and contains oxygen [7]. The oxygen arises from the interaction of sunlight on surface water molecules that break their bonds to produce hydrogen and oxygen.

### 2.2.1.2 Callisto

Callisto is another moon of Jupiter that is smaller than Ganymede, but larger than Earth's moon. The atmosphere is very thin and comprised primarily of carbon dioxide. Reference 7 suggests that Callisto is a combination of ice and rock.

Callisto's icy and rocky surface is estimated to be about 100 km thick and is heavily cratered. An ocean of salty water is presumed to reside under the surface ice. The existence of the ocean is inferred from magnetic field measurements that provide a broad range of possible depths ranging from 10 to 300 km [7]. This depth variation depends on the type and concentration of materials in the water. The interior of Callisto is presumed to be rock with a density that increases with depth.

### 2.2.1.3 Europa

Another of Jupiter's moons Europa is smaller than Earth's moon and has a diameter of about 3100 km. It has a grooved icy terrain with few craters [15]. The Galileo probe provided data that improved the understanding of Europa.

Europa is believed to have an ocean of liquid water under its icy crust that could be about 100 km in depth. A rocky mantle forms the seabed of this ocean [7]. The existence of this ocean has caused considerable speculation regarding the presence of possible life forms. At the present time, this is merely speculation without supporting evidence.

The surface of Europa is smooth that suggests a relative young origin given the dearth of craters. Its surface is also uniquely colored with interesting patterns. Neither of these unique features has been satisfactorily explained [7].

The atmosphere of Europa is limited and contains oxygen. In a manner previously addressed in the Ganymede discussion, oxygen is liberated from the interaction of sunlight on surface water molecules.

#### *2.2.1.4 Io*

Io is about the same size as the Moon, is the innermost of Jupiter's satellites, and has one of the largest densities of any Solar System moon [15]. It has some of the most active volcanoes in the Solar System. These volcanic eruptions are likely attributed to gravitational tidal stresses caused by Jupiter's mass. Io's orbit and its relationship to Europa, Callisto, and Ganymede also contribute to the tidal stresses [7].

Voyager 1 observed a number of Io's volcanic eruptions. Io's volcanoes vigorously eject sulfur compounds into the atmosphere. A portion of this material produces a ring around Jupiter. The degree of volcanic activity creates a relatively smooth surface with few observable craters [7].

Io has a number of unique surface features that are enhanced by the volcanism. It has liquid sulfur lakes, mountains, and sulfur lava flows. Io's unique and varied coloration is attributed to the various sulfur compounds. In contract to Callisto and Europa, Io has minimal water.

The Galileo probe provided data regarding Io's core. The core is likely composed of iron with a diameter exceeding 900 km.

#### *2.2.1.5 Titan*

Saturn's moon Titan with a diameter of about 5200 km is the second largest moon in the Solar System [15]. Titan is larger than Earth's moon. It has a heavy atmosphere with thick clouds. The atmosphere is composed primarily of nitrogen with some methane and trace gases with a pressure of about 1.5 atm [7, 15]. The surface temperature is about 93 K [7].

The Cassini spacecraft and its Huygens lander were able to penetrate Titan's thick clouds. Additional data were provided by the Hubble Space Telescope. The surface appears to be relatively flat, and impact craters could be filled with hydrocarbon precipitation [7].

NASA's Cassini probe observed indications of erosion on Titan caused by a flowing liquid. It is likely that the liquid is methane and there are liquid methane lakes near Titan's poles [11]. The European probe Huygens was released from Cassini, traversed its dense atmosphere, and landed on the surface of Titan. Huygens data suggests that Titan has an orange color, and its atmosphere contains clouds that resemble smog [11].

#### *2.2.1.6 Triton*

Neptune's largest moon Triton has a diameter of about 2700 km and it orbits the planet in a retrograde manner [15]. It is possible that Triton's orbital characteristics are a consequence of its gravitational capture by Neptune [7]. Voyager 2 extended knowledge of Triton and its characteristics.

Triton has a thin nitrogen atmosphere, records a surface temperature of about 33 K, and exhibits active surface features [7]. It has a bright surface that may have been impacted by geyser-like eruptions that eject liquid nitrogen into the atmosphere [11]. Solar radiation is attributed to be the driving force for these geysers [7].

The surface appears to be relatively young since it has minimal surface craters. Triton is about 25% water ice with the remaining material being rocky in composition [7].

NASA's Voyager 2 probe and the Hubble Space Telescope provided significant data regarding Neptune and its moons [7]. Additional probes and space missions are required to better characterize Neptune and its moons.

#### *2.2.1.7 Moon*

The Moon is Earth's only satellite. It has a diameter about 25% of Earth's, a mass about 0.01 times the mass of Earth, and its surface gravity is about one sixth g (~1.6 $m/s^2$). The Moon's orbit is tidally locked, and the same side faces the Earth [7].

The second densest planetary satellite in the Solar System is Earth's moon [7]. It has a thin crust, thick mantle, and an iron core that is estimated to be less than 400 km in diameter. The Moon's surface is inactive with no apparent volcanic or tectonic activity [7]. In addition, there is a minimal magnetic field approximately 100 times weaker than Earth's [7].

The Moon's surface contains numerous craters, dark areas that appear to be flat (*mare*), and basins that were created by impacts that subsequently filled with lava that has solidified [7]. The bright regions are mostly mountains and other elevated areas. The Moon's far side has minimal *mare* areas [7]. Its surface has a thin layer of fine particles of powered surface rock (regolith). NASA's Apollo missions provided considerable Lunar data including the return of sample materials to Earth.

The Moon's has essentially no atmosphere. Some material (e.g., radon gas) is released from the surface through radioactive decay. Other material is generated through micrometeorite impact and sunlight interactions with surface materials. However, much of the atmospheric constituents are swept from the Moon by the action of the Solar wind [7].

### *2.2.2 Asteroid Belt*

Several thousand bodies (asteroids) reside between 2 and 3.5 AU between the orbits of Mars and Jupiter [15]. This region is known as the Asteroid Belt. The largest of these is Ceres that contains about one third of the Asteroid Belt mass, and has a radius of about 500 km that is smaller than the Solar satellites summarized in **Table 4** [15]. Solar probes have investigated the properties of Asteroid Belt bodies. For example, NASA's space probe Dawn orbited Vesta that is one of the largest asteroids, and also investigated Ceres [11]. Some asteroid bodies have orbits that intersect Earth's orbit and present a potential collision hazard [7].

Some Solar System planets contain small moons that have a rocky composition that is similar to the Asteroid Belt bodies. It is possible that some of these moons were once asteroids that were gravitationally captured by the planets during an early phase of Solar System evolution when orbits were less stable than the current configuration.

**Table 5** lists the largest asteroids and their physical characteristics. A portion of the data was derived from the Dawn spacecraft. The mass values are rough estimates provided by NASA [25]. Ceres is classified as a dwarf planet [7]. Vesta, Pallas, and Juno and are the second, third, and fourth heaviest asteroids, respectively.

| Asteroid | Mass (kg) | Rotation Period (h) | Orbital Period (y) | Eccentricity |
|---|---|---|---|---|
| Ceres | $9.4x10^{20}$ | 9.1 | 4.6 | 0.076 |
| Pallas | $2.1x10^{20}$ | 7.8 | 4.6 | 0.23 |
| Juno | $2.0x10^{19}$ | 7.2 | 4.4 | 0.26 |
| Vesta | $2.6x10^{20}$ | 5.3 | 3.6 | 0.089 |

**Table 5.**
*Selected Asteroid Belt data derived from Ref. [25].*

Ceres is believed to have a rocky core [7]. An icy mantle and crust surround the core. The mantle could be as thick as 120 km and contain a volume of water greater than the Earth's oceans. Ceres has a surface temperature of about 233 K [7].

### *2.2.3 Comets and Kuiper Belt Objects*

In addition to moons and Asteroid Belt Objects, comets are another group of objects that orbit the Sun. Most comets are composed of a combination of rocky material, dust, and ice. A tail comprised of escaping dust and gases often characterizes comets. These materials are generated as the comet material evaporates. This evaporation is facilitated by the radiation pressure of the Sun, and the associated Solar wind of charged particles [15].

#### *2.2.3.1 Comets*

Most comets are small bodies with diameters less than 1 km to as large as 300 km [30], and many of these bodies move in highly eccentric orbits. Comets often display bright heads and long tails due to the evaporation of ice when their orbits bring them in proximity to the Sun. These bodies are probably the remnants of planetesimals originally located in an outer region of the Solar System known as the Oort cloud that lies roughly between 3,000–100,000 AU. Comets also originate in the Kuiper Belt that lies between 30 and 1,000 AU. Their orbits can be altered by the Solar System planets particularly Jupiter [15].

Comets are often classified by their orbital period. Short- (long-) period comets have orbital periods of 200 years or less (greater than 200 years) [15].

#### *2.2.3.2 Kuiper Belt Objects*

A consideration of orbital parameters suggests that the likely source of short-period comets is the Kuiper Belt. This belt contains numerous icy bodies. A selected set of Kuiper Belt Objects is summarized in **Table 6**. The largest member of the known Kuiper Belt Objects is the dwarf planet Eris. The dwarf planets Pluto and Charon, one of its moons, are also among the largest Kuiper Belt Objects [15].

The long-period comets likely originate within the Oort Cloud that is a roughly spherically symmetric region of space. These comet nuclei represent ISM dating from the time of the formation of the Solar System [15].

**Table 6** summarizes a listing of the largest Kuiper Belt Objects. In addition to Pluto, some of these objects also have an associated moon. Eris is also similar to Pluto in terms of its composition. The reader should note that Sedna has a significantly larger orbit than many observed Kuiper Belt Objects. Given this consideration, the detection of additional Kuiper Belt Objects is possible [15]. One curious possibility, known as Planet Nine, is addressed in subsequent commentary.

| Kuiper Belt Object | Diameter (km) | Orbital Period (y) | Semimajor Axis (AU) | Eccentricity |
|---|---|---|---|---|
| 2002 AW197 | 900 | 330 | 47 | 0.13 |
| Varuna | 900 | 280 | 43 | 0.052 |
| Ixion | 1100 | 250 | 40 | 0.24 |
| Quaoar | 1200 | 290 | 44 | 0.035 |
| 2003 EL61 | 1200 | 290 | 43 | 0.18 |
| 2005 FY9 | 1300 | 310 | 46 | 0.16 |
| Charon | 1300 | 250 | 39 | 0.25 |
| Orcus | 1500 | 250 | 39 | 0.22 |
| Sedna | 1600 | 12,000 | 530 | 0.86 |
| Pluto | 2300 | 250 | 39 | 0.25 |
| Eris | 2400 | 560 | 68 | 0.44 |

[a]*Derived from Ref. [15].*

**Table 6.**
*Largest Kuiper Belt Objects.*[a]

#### *2.2.3.2.1 Pluto*

Perturbations to Neptune's orbit suggested the existence of a new planet that led to the discovery of the dwarf planet Pluto. However, Pluto's mass is too small to have perturbed Neptune's orbit. The orbital perturbation issue involving other Kuiper Belt Objects has led to the speculation of additional Solar System planets [15, 29].

Pluto is likely composed of rock, and it has an icy surface mixture of frozen nitrogen with smaller amounts of carbon monoxide and methane. As such, Pluto's surface is similar to Triton [15]. It is believed to have a rocky core. Pluto has a very thin atmosphere that includes nitrogen, methane, and carbon monoxide [15]. This composition is consistent with the sublimation of surface material [15].

Pluto has a diameter of about 2300 km and has a rotation period of about 6.4 days. It has five moons [28]. Pluto's largest moon (Charon) and its characteristics are summarized in **Table 6** [15].

#### *2.2.3.2.2 Eris*

Eris is an additional dwarf planet that is slightly larger than Pluto. It has a radius about 20% of Earth's. The internal structure of Eris is unknown [27]. Ref. [15] suggests that its surface is similar to Pluto that incorporates a composition of frozen methane, rock, and ices of various elements including nitrogen. Eris' surface temperature is about 33 K [27]. This dwarf planet also has a small moon Dysnomia that has a circular orbit with a rotational period of about 16 days.

#### *2.2.3.2.3 Planet Nine*

Periodically, predictions of additional planets beyond Pluto have been made. One of the more recent developments is an assertion by Batygin and Brown [12]. Ref. [12] suggests the existence of an additional planet with a mass of approximately 10 Earth masses, and an orbit at about 20 times farther from the Sun than Neptune. Their estimates for Planet Nine were inferred from observed orbital characteristics of a number of Kuiper Belt Objects.

Ref. [12] performed numerical simulations to fit the observed orbits. A perturbing body (Planet Nine) was incorporated in the simulation to reproduce the observed orbital characteristics of a number of Kuiper Belt Objects. The simulations [12] suggest that the object perturbing the orbits is a "primordial giant planet core that was ejected during the nebular epoch of the Solar System's evolution". At the present time, the existence of Planet Nine is unconfirmed.

### *2.2.4 Meteorites*

The collision of asteroids produces a collection of smaller objects known as meteoroids. These collisions produce trajectories governed by the gravitational interaction of Solar System bodies. When a meteoroid enters the atmosphere of a planet, it is heated by friction and becomes a meteor. If the meteor strikes the planet's surface, its surviving fragments are called meteorites [15].

Meteorite material is also produced by the fracturing of comets that are exposed to the increased heat of the Sun particularly in the vicinity of the inner Solar System planets. These occurrences often lead to showers of micrometeorites that impact a planet's atmosphere [15].

Meteorites play a key role in planetary development. By depositing chemical elements or prebiotic material, meteorites can influence the evolution of a plant during its formative years. The extent of this influence depends on their size, frequency of impact, and composition. Meteorites striking a mature planet can cause ecological harm including impacts on the climate.

## 2.3 Exoplanetary Systems

Exoplanets are planets that orbit stars outside the Solar System. Given the diversity of star types and sizes, exoplanets have a wider range of physical characteristics than the planets that inhabit the Solar System. The various types of exoplanets include the most massive or gas giant planets, intermediate mass or Neptune planets, and low mass planets that include terrestrial and ocean planets or water worlds [9, 26]. Although most exoplanets orbit their host stars, rogue exoplanets are also possible [8].

### *2.3.1 Gas Giants*

Gas giant exoplanets are similar to Jupiter and Saturn. Their composition is dominated by hydrogen and helium with smaller contributions from heavier elements and complex molecules. Their structures may include cores of rock or ice [9, 15, 26]. However, a recent publication suggests that hot Jupiters could exhibit a more diverse chemical composition [34]. For example, an analysis of HD 209458b atmospheric data suggests the presence of water, carbon monoxide, hydrogen cyanide, methane, ammonia, and acetylene [34].

Hot Jupiters are a classification of gas giants that typically reside near their star usually within 0.05 to 0.5 AU [23]. They could have formed near their host star or migrated toward the star after forming at a more distant location. Given the proximity to their host star, temperatures can exceed 2,000 K [23]. Considering their size and proximity to their host star, they are readily detected and are one of the most common types of exoplanet detected to date. A commentary of detection methods is presented in subsequent discussion.

#### *2.3.2 Neptune Class*

Exoplanets known as the Neptune class are also giant planets, but heavy elements comprise most of their mass [9, 26]. The Neptune class of exoplanets has a thick hydrogen and helium layer, but these elements are not the dominant constituents as they were in the gas giants. Within the Solar System, both Uranus and Neptune are representative of this class of planets.

Uranus and Neptune are Solar System analogues of Neptune class exoplanets. These systems have a characteristic blue color. They are also known as ice giants because many models suggest that the bulk of the planet's mass resides within a sea that probably is comprised of ammonia, methane, and water [9]. However, there is likely a significant diversity in the composition of Neptune class exoplanets.

#### *2.3.3 Terrestrials*

Terrestrials are exoplanets that are similar in structure to the inner planets of the Solar System including Mercury, Venus, Earth, and Mars. These exoplanets have compositions that are dominated by elements including carbon, oxygen, magnesium, silicon, and iron [9, 26]. Terrestrials are interesting because these bodies present the possibility of finding a planet similar to Earth that could support life.

A sub-classification of terrestrials is Super-Earths. Super-Earths are rocky planets that have a mass greater than the Earth, but usually defined to be less than 10 Earth masses. Given their mass, some Super-Earths may be similar to Solar System planets Neptune and Uranus [23].

Exo-Earths are terrestrial exoplanets that have similarities to Earth in terms of their mass, radius, and temperature. Their orbits would reside within the habitable zone where liquid would water could exist [23].

Chthonian Planets are a proposed class of exoplanets that began as gas giants. During subsequent evolution, their orbits were altered to bring these planets in proximity to their host star. This proximity caused their atmosphere to be removed with only the rocky core remnant remaining. Given their similarity, some Super-Earths could be Chthonian Planets [23].

#### *2.3.4 Water worlds*

Exoplanets, referred to as ocean planets or water worlds, are dominantly comprised of water. Computer models suggest these exoplanets could be created by ISM enriched in icy material. As a candidate water world migrated toward the host star, the ice melted, and covered the planet in an ocean [23, 26].

## 3. Detection methods

A number of detection methods have been utilized to observe exoplanets. These include, but are not limited to, transient methods, direct detection, and radial velocity measurements. Each of these three basic methods is addressed in subsequent discussion. If concurrent radial velocity and transient methods measurements are available, this combination can be used to determine the planet's mass, radius, and density [23, 33].

Ref. [32] notes that in order to estimate an exoplanet's mass, the mass of the host star must be determined. The host star's mass estimate is based on its spectral type.

An exoplanet's mass can then be estimated by measuring its effects on the motion of the host star. Included in these effects is stellar wobble that periodically red shifts and blue shifts its emitted spectrum of light. Measuring these shifts as a function of time permits a determination of the orbital period. When combined with the host star's mass, the planet's orbital trajectory and velocity can be estimated. Once the star's velocity is known as a function of its wobble, then the exoplanet's mass is determined. The star's velocity is not zero because both the planet and host star orbit around their center-of-mass.

### 3.1 Transit method

The transient method is the detection approach used by a number of devices including the Kepler Space Telescope. Using this method, an observer or instrument detects the decrease in radiation intensity when an exoplanet transits its host star. Periodic measurements create an intensity vs. time plot or a light curve. The light curve has a characteristic shape. It has a constant intensity until the exoplanet initiates its transit. As the exoplanet begins to cover a portion of the star, it blocks some the star's surface. When this occurs, the intensity is reduced and decreases to a reduced constant value. As the exoplanet blockage terminates, the intensity returns to its initial value before any light was blocked. For an exoplanet, this cycle repeats as the exoplanet periodically orbits and transits its host star. A dip or decrease in intensity indicates that the exoplanet is passing between the observer and its host star. If this intensity curve pattern recurs on a periodic basis, then an exoplanet has likely been observed [23].

The drop (D) in intensity (I) between the equilibrium or maximum value and the minimum value as the exoplanet is transiting its host star is given by the first order relationship:

$$D = \frac{I_{minimum}}{I_{maximum}} = \left(\frac{r}{R}\right)^2 \tag{1}$$

where r is the exoplanet radius and R is the host star radius. Eq. (1) assumes the star and exoplanet are both spherical. Using this assumption, the volume (V) of the exoplanet can be obtained from Eq. (1):

$$V = \frac{4}{3}\pi r^3 \tag{2}$$

Since the exoplanet is much smaller that its host star, the decrease in intensity is relatively small and dependent on the relative size of the planet and star. As an example, consider an exoplanet with a radius that 10% of its host star's. Using Eq. (1), the light will only dim by 1% if both bodies are spherical.

**Figure 2** illustrates the intensity profile for a transiting exoplanet across its host star. The figure assumes a large exoplanet and the intensity decrease is not normally so dramatic.

### 3.2 Direct detection

Direct detection entails the viewing or imaging of the exoplanet [7, 23]. This differs from the transit method because an explicit observation of the exoplanet

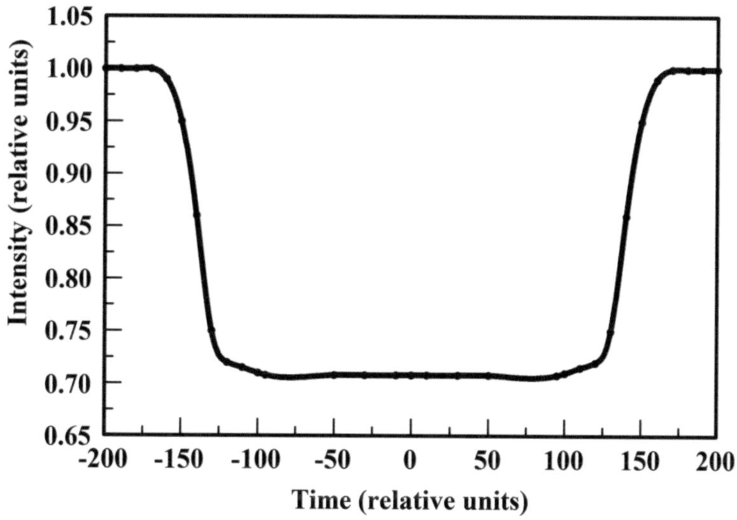


**Figure 2.**
*Intensity profile for a large exoplanet traversing its host star.*

occurs. Although direct detection is desirable, there are two basic complications when implementing this method.

First, the star's intensity is orders of magnitude larger than the reflected and internally generated light from an exoplanet. This difference in intensity must be mitigated for the exoplanet to be observed. Second, the detection instrumentation must have the necessary angular resolution to distinguish the exoplanet from its host star. If this is not accomplished, the image will have insufficient resolution to reveal the exoplanet.

These issues create difficulties in the detection of exoplanets occurring in a habitable zone where liquid water exists. Terrestrial exoplanets that are close to a parent star such as a red giant would preclude imaging the surface topography. This proximity minimizes the direct observation of details including continents and oceans. Attention could be focused on observing atmospheric constituents to ascertain their elemental composition including spectroscopic evidence of any possible signatures of biological activity.

### 3.3 Radial velocity

The radial velocity approach entails measurement of the Doppler shift of light from a host star. As an exoplanet orbits its host star, it exerts a gravitational force on the star and causes a shift in its radial velocity as it moves toward and away from the observer. This radial velocity shift results in a wavelength frequency that periodically oscillates between a red shift and a blue shift. When this periodic oscillating frequency is observed, the host star likely has one or more exoplanets in its orbital system [7, 15]. The oscillating effect of red shifts and blue shifts is illustrated in **Figure 3**.

### 3.4 Trappist-1 exoplanetary system

The numerous exoplanets discovered to date are a testament to improving detection methods. Although a complete description of these systems is not possible in this chapter, it is possible to illustrate one of these systems that has several interesting characteristics. Accordingly, this section addresses the Trappist-1 system that

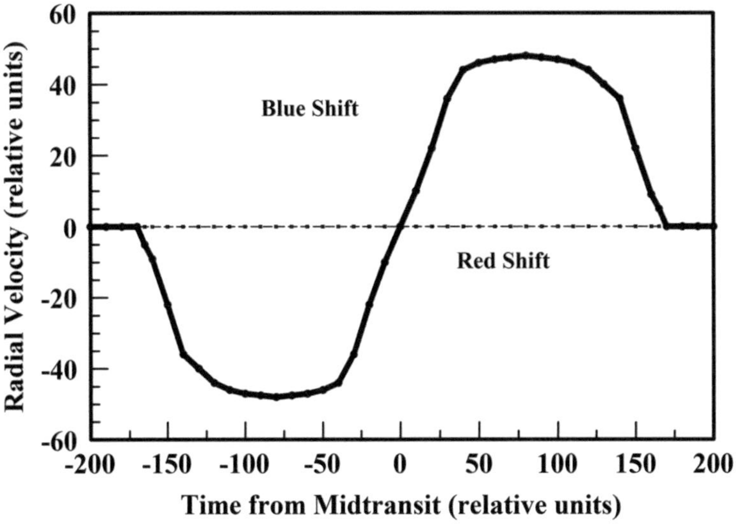


**Figure 3.**
*Radial velocity profile for a transiting exoplanet. Positive (negative) radial velocities result in a light profile that is blue (red) shifted.*

| Exoplanet[a] | $m(M_\oplus)$ | $r(R_\oplus)$ | $\rho(\rho_\oplus)$ | Surface gravity (g) | Semi-major axis (AU) | Orbital Period[b] (d) |
|---|---|---|---|---|---|---|
| Trappist-1b | 1.02 | 1.12 | 0.73 | 0.81 | 0.012 | 1.51 |
| Trappist-1c | 1.16 | 1.10 | 0.88 | 0.97 | 0.016 | 2.42 |
| Trappist-1d | 0.30 | 0.78 | 0.62 | 0.48 | 0.022 | 4.05 |
| Trappist-1e | 0.77 | 0.91 | 1.02 | 0.93 | 0.029 | 6.10 |
| Trappist-1f | 0.93 | 1.05 | 0.82 | 0.85 | 0.039 | 9.21 |
| Trappist-1 g | 1.15 | 1.15 | 0.76 | 0.87 | 0.047 | 12.4 |
| Trappist-1 h | 0.33 | 0.77 | 0.72 | 0.56 | 0.062 | ~20 |

[a]*Derived from Ref. [17].*
[b]*Derived from Ref. [16].*

**Table 7.**
*Selected characteristics of Trappist-1 exoplanets.*[a,b]

contains seven potential Earth-like exoplanets. Trappist-1 is an ultra-cool red dwarf star with a radius that is somewhat larger than Jupiter's, but has a mass of about 84 times Jupiter's [13, 16]. It has a surface temperature of about 2600 K [16], which partially explains the nature of the habitable semi-major axes and rotation periods noted in **Table** 7.

In 2017, NASA announced the observation that seven rocky exoplanets similar in size to Earth were discovered orbiting the host star Trappist-1 [16]. These planets resided in the habitable zone and had the potential for the existence of liquid water on their surfaces. Trappist-1 is about 40 light years from Earth, and this proximity creates the possibility that its planetary systems could be imaged with future generations of telescopes [16]. Further investigation could reveal the existence of atmospheric constituents that would indicate the possible presence of life forms.

**Table** 7 summarizes selected details of the Trappist-1 exoplanetary system [16, 17]. These exoplanets have nearly circular orbits, and orbit in proximity to their host star. The Trappist-1 exoplanets are in the range of 0.3–1.2 Earth masses with radii of 0.8–1.2 Earth radii. These rocky worlds have densities between 0.6 and 1.0 Earth densities. Their surface gravity values are also similar to Earth's (0.5–1.0 g).

Using simulations, Ref. [17] reached several conclusions regarding the nature of the Trappist-1 exoplanetary system. Three of these most applicable to this chapter are noted.

First, Trappist-1c and -1e are likely to possess interiors that are mostly rocky in nature. Second, Trappist-1b, −1d, −1f, −1 g, and -1 h likely have a thick atmosphere, oceans, or ice cover. Third, Trappist-1d, −1f, −1 g, and -1 h are unlikely to have an enriched carbon dioxide atmosphere above a bare core assuming these planets have an Earth-like composition.

## 4. Space probes

A variety of probes have investigated Solar System planets as well as exoplanets. These devices are growing in capability, and future probes could have the capability to reveal significant details regarding the exoplanetary structure and atmospheric composition of the increasing number of observed exoplanetary systems.

### 4.1 Solar system probes

There have been numerous scientific probes launched primarily by the United States, the European Union, and Russia/Soviet Union since the late 1950s [19]. This list of participating countries has expanded to include many more nations as launch capabilities extend to additional nations. The probes have included a variety of purposes and target planets.

Missions include flybys of a planet, moon, or other space objects; orbiting these bodies; atmospheric entry; impact with the space object; and soft-landing on the surface. Target space bodies include all Solar System planets, the Moon, dwarf planets, asteroids, moons of planets, asteroids, and comets. Probes have also been launched into Solar orbits.

The probes have expanded our knowledge of planetary systems including their masses, atmospheric composition, surface characteristics, and temperature and pressure profiles. Additional studies focused on major moons as well as other orbiting space objects including comets and asteroids.

Ref. [19] provides a detailed historical listing of space probes during 1958–2016. Since 2016, the number of probes and their sophistication has improved. Although, Mars and the Moon are popular destinations, recent data, suggesting the appearance of phosphine in the atmosphere of Venus, has increased focus of that planet [21]. The number and scale of future probes will reveal additional information regarding these planets. For this reason, this chapter has been written in a general manner and does not address specific probe objectives or data. However, some of the chapters of this book illustrate selected probes and their capabilities.

### 4.2 Exoplanet probes

The Exoplanet Exploration Program's roadmap of NASA's exoplanet missions provides a summary of existing and planned probes [20]. NASA has a number of ongoing exoplanet and planned probes. Current instrumentation includes the Hubble, Spitzer, and Kepler/K2 Space Telescopes. These instruments have significantly improved knowledge of exoplanetary systems though the discovery and characterization of transiting systems. Another device, the Transiting Exoplanet Survey Satellite (TESS) launched in 2018, adds an additional tool to expand the knowledge of exoplanetary systems. The James Webb Space Telescope (JWST), having a projected late 2021 launch date, includes mid-infrared transit spectroscopy, and has the capability to directly image massive exoplanets. In addition, the Nancy Grace Roman Space Telescope tentatively planned for launch about 2025 includes infrared capability that could facilitate direct exoplanet imaging.

Additional missions are possible if oversight groups recommend their viability, and funding becomes available. These include the Large Ultraviolet Optical Infrared Surveyor (LUVOIR) and the Habitable Exoplanet Imaging Mission (HabEx) that have the potential to directly image and characterize Earth-like exoplanets. These devices also have the capability to detect the spectra of molecular forms including water and oxygen. The Origins Space Telescope (OST) is a conceptual infrared mission that is intended to measure the atmosphere of exoplanets. These additional space missions have the capable of observing the necessary characteristics for life to exist on exoplanets.

## 5. Conclusions

Observations of Solar System planets have been ongoing for centuries. The expanding use of probes has significantly improved knowledge of the major planets, dwarf planets, moons, the Asteroid Belt, and Kuiper Belt Objects. Although the Solar System characteristics are relatively well known, probes have not extensively studies all the planets and dwarf planets at the same level of detail. Accordingly, Venus, Neptune, Uranus, and Pluto warrant additional investigation. The dynamics of Solar System planets are also evolving as evidenced by the periodic suggestion of additional planets including the recent calculations regarding the existence of Planet Nine.

Although exoplanets are a relatively new area of study, the field is rapidly advancing with thousands of exoplanets currently catalogued. Much is to be learned about these worlds and their characteristics, and a number of new and planned probes will add to that knowledge. Although manned planetary missions are on the horizon, exoplanet investigations will require remote approaches for the foreseeable future.

## Author details

Joseph Bevelacqua
Bevelacqua Resources, Richland, WA, United States of America

*Address all correspondence to: bevelresou@aol.com

## References

[1] Anderson, H. L. (Editor). AIP 50th Anniversary Physics Vade Mecum. New York: American Institute of Physics; 1981. 330 p.

[2] Mayor, M., Queloz, D. A Jupiter-mass companion to a solar-type star. Nature. 1995;**378**:355-359.

[3] Hartmann, W. K. *Moons and Planets*. 5th ed. Belmont, CA: Thomson Brooks/Cole: 2004. 456 p.

[4] Bevelacqua J. J. Health Physics in the 21st Century. Weinheim: Wiley VCH; 2008. 562 p.

[5] Bevelacqua J. J. Contemporary Health Physics: Problems and Solutions. 2nd ed. Weinheim: Wiley VCH; 2009. 722 p.

[6] Bevelacqua J. J. Basic Health Physics: Problems and Solutions. 2nd ed. Weinheim: Wiley VCH; 2010. 768 p.

[7] Barnard, L. et al., eds. Astronomica. North Narrabeen, NSW, Australia: Millennium House; 2011. 524 p.

[8] Bevelacqua, J. J. Can Extra Dimensions be Detected from Planetary Orbital Observations?. Physics Essays. 2013; 26: 381 – 387.

[9] Spiegel, D. S., Fortney, J. J., Sotin, C. Structure of Exoplanets. PNAS. 2014; **111:** 12622 - 12627.

[10] de Pater, I., Lissauer, J. J. Planetary Sciences. 2nd ed. Cambridge: Cambridge University Press; 2015. 688 p.

[11] Ridpath, I. Exploring the Solar System. 2015. Available from: www.ianridpath.com/solarsystem.pdf [Accessed 2021-02-14].

[12] Batygin, K., Brown, M. E. Evidence for a Distant Giant Planet in the Solar System. The Astronomical Journal. 2016; 151:22:1-12.

[13] Gillon, M. et al. Temperate Earth-sized planets transiting a nearby ultracool dwarf star. Nature. 2016; 553: 221-224.

[14] Bevelacqua, J. J. Radiation Protection Consequences of the Emerging Space Tourism Industry. J J Earth Science. 2017; 1 (003): 1 – 11.

[15] Carroll, B. W., Ostlie, D. A. An Introduction to Modern Astrophysics. 2nd ed. Cambridge: Cambridge University Press; 2017. 1278 p.

[16] Exoplanets: The Discovery of Seven Rocky Earth-Sized Planets Around the Star TRAPPIST-1. 2017. https://www.starrynight.com/TRAPPIST-1/PDF-ePub/Exoplanets%20-%20The%20Seven%20Earth-like%20Worlds%20of%20TRAPPIST-1%20.pdf. Available from: [Accessed 2021-02-21]

[17] Grimm, S. L., et al. The nature of the TRAPPIST-1 exoplanets. Astronomy and Astrophysics. 2018; 613 A68: 1-21.

[18] Mortazavi, S. M. J., Bevelacqua, J.J., Fornalski, K. W., Waligorski, M., Welsh, J., Doss, M. Comments on "Space: The Final Frontier—Research Relevant to Mars". Health Physics Journal. 2018; 114: 344 – 345.

[19] Siddiqi, A. A. Beyond Earth-A Chronicle of Deep Space Exploration, 1958 – 2016. Washington, DC: National Aeronautics and Space Administration; 2018. p393.

[20] NASA/JPL/Caltech. The Exoplanet Exploration Program Technology Update, 2019. Pasadena, CA: California Institute of Technology. exoplanets.nasa.gov › 146_20200103-1335_Siegler_Technology_Exopag21. [Accessed 2021-02-02].

[21] Greaves, J. S. et al. Phosphine gas in the cloud decks on Venus. Nature

Astronomy. 2020. https: doi.org/10.1038/541550-020-1174-4.

[22] den Hond, B. A. Field Guide to the Magnetic Solar System, EOS. 2021; 102: 36:41.

[23] Exoplanets. Available from www.astronomy.nmsu.edu. [Accessed 2021-02-02].

[24] Lichtenberg, T., et al. Bifurcation of planetary building blocks during Solar System formation. Nature. 2021; 371; 365:370.

[25] NASA Asteroid Fact Sheet. 2021. Available from: https://nssdc.gsfc.nasa.gov/planetary/factsheet/asteroidfact.html [Accessed 2021-02-17].

[26] NASA Exoplanet Archive. 2021. Available from: https://exoplanetarchive.ipac.caltech.edu/ [Accessed: 2021-01-28].

[27] NASA Solar System Exploration - Eris in Depth. 2021. Available from: https://solarsystem.nasa.gov/planets/dwarf-planets/eris/in-depth/. [Accessed 2021-02-17].

[28] NASA Solar System Exploration - Moons. 2021. Available from: https://solarsystem.nasa.gov/moons/overview/ [Accessed: 2021-01-27].

[29] NASA Solar System Exploration – Pluto Dwarf Planet. 2021. Available from: https://solarsystem.nasa.gov/planets/dwarf-planets/pluto/overview/. [Accessed 2021-02-25].

[30] Roundtree-Brown, M. Ten Important Comet Facts-NASA. Available from: https://deepimpact.astro.umd.edu/educ/CometFacts.html. [Acessed 2021-02-28].

[31] University of Colorado. Jupiter: The Planet, Satellites and Magnetosphere. Available from: https://lasp.colorado.edu/home/mop/bibliographies/jupiter-book-chapters/. [Accessed: 2021-02-23].

[32] Wright, J., Blake, C. How We'll Find another Earth. IEEE Spectrum. 2021: 03.21: 22 – 28.

[33] Trifonov, T. et al. A nearby transiting rocky exoplanet that is suitable for atmospheric investigation. Science. 2021: 371; 1038 – 1041.

[34] Giacobbe, P. et al. Five carbon- and nitrogen-bearing species in a hot giant planet's atmosphere. Nature. 2021:592; 205 – 208.

[35] NASA Solar System Exploration – Our Solar Syatem. 2021. Available from: https://solarsystem.nasa.gov/resources/490/our-solar-system/?category=solar-system_our-solar-system. [Accessed 2021-05-14].

Section 2

# The Sun

Chapter 2

# Millennial Oscillations of Solar Irradiance and Magnetic Field in 600–2600

*Valentina Zharkova*

## Abstract

Daily ephemeris of Sun-Earth distances in two millennia (600–2600) showed significant decreases in February–June by up to 0.005 au in millennium M1 (600–1600) and 0.011au in millennium M2 (1600–2600). The Earth's aphelion in M2 is shorter because shifted towards mid-July and perihelion longer because shifted to mid-January naturally explaining two-millennial variations (Hallstatt's cycle) of the baseline solar magnetic field measured from Earth. The S-E distance variations are shown imposed by shifts of Sun's position towards the spring equinox imposed by the gravitation of large planets, or solar inertial motion (SIM). Daily variations of total solar irradiance (TSI) calculated with these S-E distances revealed TSI increases in February–June by up to 10–12 $W/m^2$ in M1 and 14–18 $W/m^2$ in M2. There is also positive imbalance detected in the annual TSI magnitudes deposited to Earth in millennium M2 compared to millennium M1: up to 1.3 $W/m^2$, for monthly, and up to 20–25 $W/m^2$ for daily TSI magnitudes. This imbalance confirms an ascending phase of the current TSI (Hallstatt's) cycle in M2. The consequences for terrestrial atmosphere of this additional solar forcing induced by the annual TSI imbalances are evaluated. The implications of extra solar forcing for two modern grand solar minima in M2 are also discussed.

**Keywords:** Sun: solar activity, Sun: magnetic field, gravitation, solar inertial motion, solar irradiance

## 1. Introduction

Solar activity is usually classified by the numbers of sunspots appearing on the solar surface as locations of magnetic loops generated by electro-magnetic solar dynamo in the solar interior [1] with the number of sunspots on the solar surface to change periodically over an eleven-year cycle [2, 3]. Babcock [4] found that a solar background magnetic field (SBMF) surrounding sunspots has the polarities opposite to the leading sunspot polarities, and these are changing also periodically every 11 years.

The magnetic polarities of SBMF and leading polarities of sunspots are shown to be in anti-phase, e.g. having opposite polarities, as found by comparing solar background and sunspot magnetic fields for cycle 21 [5], and 23 [6]. Furthermore, the SBMF was found to be the leading force defining timing and locations of sunspot occurrences on the solar surface and migration to the equator accounting for north–south asymmetry of solar activity [6]. This investigation highlighted the important role of SBMF in generation of sunspots by dynamo actions and their appearance on the solar surface, thus, defining the solar activity. The link between SBMF (poloidal field) and sunspot (toroidal) magnetic field defines the action of the solar dynamo [1] and it would be beneficial to link a proxy of solar activity to the solar background magnetic field as it is systematically measured in the past 45 years.

Zharkova et al. [7] explored the solar background magnetic field and found the eigen values of the own oscillations of the Sun, by applying Principal Component Analysis (PCA) to the low-resolution full disk magnetograms captured in cycles 21–23 by the Wilcox Solar Observatory. This approach allowed authors to replace a complex magnetic field seen on the solar surface, the photosphere, with the separate wave components, eight plus eigen vectors, which appeared in pairs [8]. The pair of the two principal components (PCs) are the strongest waves of solar magnetic oscillations covering about 67% of the data by standard deviation, with the nearly-equal largest eigen-values, which oscillate with not equal periods of about 11 years [9, 10]. The PCs are shown to be two magnetic waves generated by the dipole magnetic sources produced by the double solar dynamo action [10, 11] in the inner and outer layers of the solar interior [12].

These waves start in the opposite hemispheres while travelling with an increasing phase shift to the Northern hemisphere (in odd cycles) and to the Southern hemisphere (in even cycles) [7, 10]. The summary curve of these two waves is found close to the averaged sunspot numbers, which define the current solar activity index [9, 10]. This summary curve of solar magnetic waves is proposed as a new proxy of solar activity, which allows us to predict solar activity on any timescale and also to add a magnetic polarity of the background field for a given cycle, known to be opposite to leading polarity of sunspots [6]. The maximum of solar activity for a given cycle (or double maximum for the double waves with a larger phase shift) occurs at the times when each of the waves approaches its maximum, so that at the equal amplitudes the two waves can have a resonant interaction, naturally accounting for often-reported North–South asymmetry of solar activity [6, 13–15].

In order to test further predictions of solar activity with the summary curve of two magnetic waves generated by double dynamo in the Sun in two layers, the summary curve was extended using the mathematical formula from the current time forwards to 3200 and backwards to 1200 [10] (see **Figure 1**). This led to a discovery of grand solar cycles (GSCs) of solar activity with a duration of 350–400 years, evidently caused by the interference (beating effect) of the two magnetic waves with close but not equal frequencies [10]. There were far fewer sunspots seen during some periods, for example, during the Dalton minimum (1790–1820), and practically none during the period known as the Maunder minimum

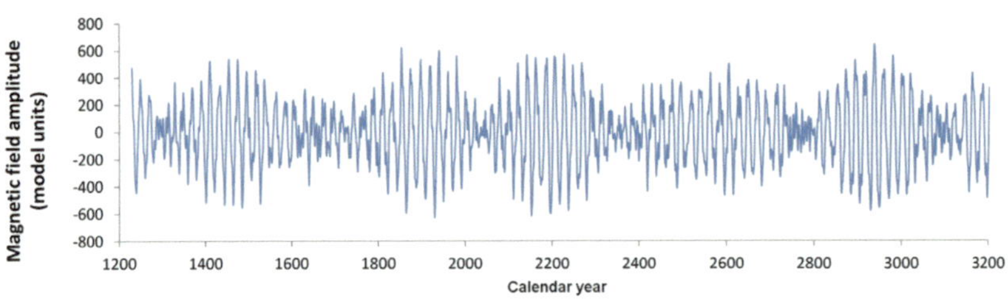


**Figure 1.**
*The summary curve (in arbitrary units) of solar activity calculated for 1200 to 3200 years from the 'historical' period (1976–2008, cycles 21–23). Positive magnitudes of the summary curve represent northern magnetic polarity while the negative ones - southern magnetic polarity. Reproduced from the data of Zharkova et al. [10].*

(1645–1715) [16]. Such dramatic reductions in solar activity, which are longer than a single eleven-year sunspot cycle, are known as grand solar minima (GSMs).

The timings of previous GSMs are found to closely fit the Maunder minimum (1645–1715) [16] and Wolf minimum (1280–1350), and to predict the two upcoming modern GSMs (2020–2053 and 2375–2415). Furthermore, by extrapolating the summary curve backwards to 1000 BC the further GSMs are fit by the curve: Oort's (1040–1080), Homeric (780–710 BC) and many others [17]. This restoration of summary curve [17] clearly gives a better accuracy of solar activity definition, in comparison with the prediction of sunspot activity restored from the past TSI derived with carbon-14 isotope dating [18].

Then, as the next step, Zharkova and co-authors [19] derived the magnitudes of a baseline (zero-line) magnetic field for each 22-year sets and discovered rather rigid periodic variations of this baseline magnetic field with a period of about 2000–2100 years. This period resembles the period of 2200 years of Hallstatt's cycle reported from the restoration of solar irradiance in the Holocene [20–22]. It is rather difficult to find any mechanism in the solar interior that can account for much weaker and longer oscillations of the baseline of magnetic field. This led us to look for a some kind of periodic forcing linked to the orbital motion of the planets.

Jose [23] first suggested that solar activity on a longer timescale can be affected by the motion of large planets of the solar system. This suggestion was later developed by many researchers (see for example [24–27]) who found that the Sun, as a central star of the solar system, is a subject to the inertial motion around the center-of mass, or barycentre, of the solar system induced by the motions of the other planets (mostly large planets, e.g. Neptune, Jupiter, Saturn and Uranus).

Solar inertial motion (SIM) is the motion of the Sun around this barycenter of the solar system inside the circle with a diameter of about $\Delta = 4.3 R_{Sun}$, or $\Delta = 4.3 \times 6.95 \cdot 10^5 = 2.9885 \cdot 10^6$ km, where $R_{Sun}$ is a solar radius as shown in **Figures 4** and **5** in [19] reproduced from [27, 28]. This schematic drawing (see Fig. 4 in [19]) illustrates sudden shifts in the Sun from the location in the ellipse focus, where it is supposed to reside by Kepler's laws, because the Sun travels in an epitrochiod-shaped orbits about the center-of mass (barycentre) of the solar system.

The SIM orbits are induced the tri-fall positions of large planets achieved for different planet configurations changing approximately within different periods of 370 or 2200–2400 years related to the planet positions and their rotation around the Sun [29, 30]. Hence, a joint effect of the orbital effects introduced by the combined motion of the Earth on the orbit and the Sun about the barycentre of the solar system can be the important factors in defining the observed long timescale variations of solar irradiance at the Earth and terrestrial temperature, which has not been explored yet, despite the Sun is the main source of energy of all the planets (**Table 1**).

Solar irradiance is accepted to be one of the important factors defining temperature variations on the Earth and other planets as it is the main source of the energy

| Authors | Maunder minimum | 2000–2012 |
|---|---|---|
| Lean et al.1995 [31] | 1363 | 1366 |
| Steinhilber et al., 2012 [22] | 1364 | 1366 |
| Shirley et al., 1990 [26] | — | 1370 |
| Wolff and Hickey, 1987 [32] | — | 1371 |
| Lee et ak, 1985 [33] | — | 1372 |

**Table 1.**
*The solar irradiance in $W/m^2$ restored and measured since Maunder minimum.*

for all planets. During the Maunder minimum, solar activity was significantly reduced for six solar cycles of 11 years and so was the terrestrial temperature in the Northern hemisphere. This was considered to be a result of a reduction of solar irradiance during the Maunder Minimum.

More recent reconstruction of the cycle-averaged solar total irradiance back to 1610 suggests that since the end of the Maunder minimum in 1710 until 2010 there was the increase of the irradiance by a magnitude about $1 - 1.5\ W/m^2$ [34]. This increase is correlated somehow with the increase of the baseline terrestrial temperature since the Maunder minimum (e.g. recovering from the little ice age) [35]. Although, it is not clear yet if this trend in variations of the terrestrial temperature and solar irradiance is caused by the increased solar activity itself, which, in fact, started to decrease in the past decades, or by some other factors of the solar-terrestrial interaction and by human activities, or by the combination of all the three factors.

In the current chapter we analyse the observational variations of Sun-Earth distances derived from the published ephemeris in the two millennia 600–2600 and relate them to the variations of solar irradiance at the Earth and explore their possible links oscillations of the baseline solar magnetic field and with the reported planetary motions.

## 2. Observed millennial oscillations of solar irradiance and baseline solar magnetic field

### 2.1 Millennial oscillations of solar irradiance

Reconstruction of cycle-averaged total solar irradiance (TSI) back to 1610 suggests an increase of the solar irradiance by a value of about $3\ W/m^2$ (see **Figure 2**) [31, 37], or about 0.22% of the total solar irradiance since the end of the Maunder minimum (see **Figure 2**, left plot).

The space observations in 80s of the total solar irradiance obtained by NIMBUS 7 instruments show pretty wide range of magnitudes varying up to 1370 $W/m^2$ [26], to 1371 $W/m^2$ [32] or 1372 $W/m^2$ [33]. The wide variety of the measured magnitudes of solar irradiance indicates that this physical parameter from the Sun is not as constant as many researchers assume. Although, these changes of solar irradiance from the MM until present times are, in general, small, compared to the tens of watts occurring during seasonal and latitude differences, which may have a noticeable impact on the Earth temperature.

Note, we do not include in this comparison the most recent restorations of the solar irradiance [34, 38], who considered the re-normalised solar irradiance after Maunder minimum and used a magnetic flux transport model with strongly averaged past solar magnetic fields, which make rather difficult to compare these magnitudes of solar irradiance with the non-normalised early observations.

The variations of the solar irradiance recovered for the Holocene from the variations of the carbon $^{14}C$ isotope abundance in the terrestrial biomass [39] (see **Figure 3**, top plot), demonstrate weak oscillations with a period of about 2200 years, or Hallstatt's cycle [20, 21], which are imposed onto the longer-term (16-20 K years) orbital oscillations (possibly, one of Milankovich cycles) [40, 41]. The solar irradiance oscillations restored over the past 12 000 years [20, 42] were also tested with the wavelet transform spectral analysis, which clearly demonstrate the similar period of 2200 years [22] or up to 2400 years [43]. These baseline

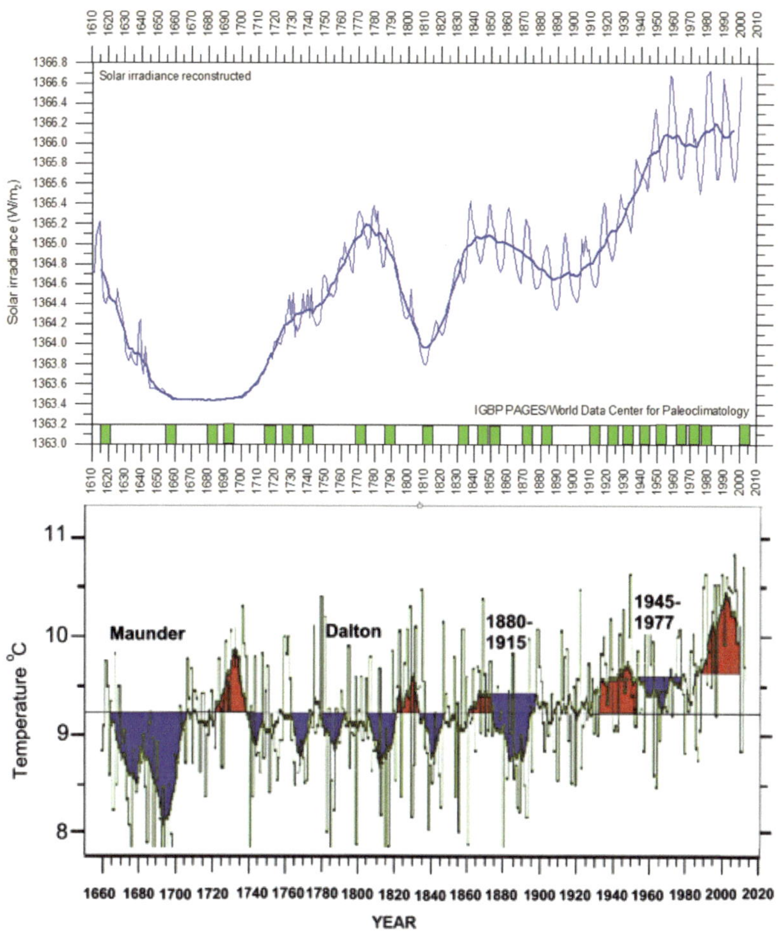


**Figure 2.**
*The variations of solar irradiance (left) [31] and terrestrial temperature (right) [36] recovered from the Maunder minimum, which demonstrates a significant drop of the solar irradiance and terrestrial temperature during the previous GSM, Maunder minimum (see the text for details).*

oscillation periods are very close to the 2200 year period called Hallstatt's cycle reported from the other observations of the Sun and planets [21, 22, 24, 44].

### 2.2 Millennial oscillations of the baseline magnetic field

Recently, Zharkova et al. [19] reported the similar millennial oscillations of the baseline (zero-line) of the solar background magnetic field (SBMF) calculated from the summary curve obtained with Principal Component Analysis (PCA) [10]. The baseline magnetic field is defined from filtering out large-scale 22 year oscillations, or finding the mean point between two 11 year cycles for the expanded summary curve of 120 thousand years. As result, we detect weak two millennial oscillations of the SBMF baseline with a period of 2000 $\pm$ 95 years [19] shown by the navy curve in **Figure 3** (bottom plot). Although, the scale of these baseline oscillations is much smaller (ranging from $-10$ to 10) than the 11 year magnetic field variations of the summary curve (ranging in $-400,400$) that is shown in **Figure 3** (bottom plot) for the redacted summary curve (cyan curve) calculated backwards between 70 and 90 thousand years [19]. Note, the summary curve presented by cyan curve in **Figure 3** (bottom plot) has different appearance from that in **Figure 1** (top plot) [19]

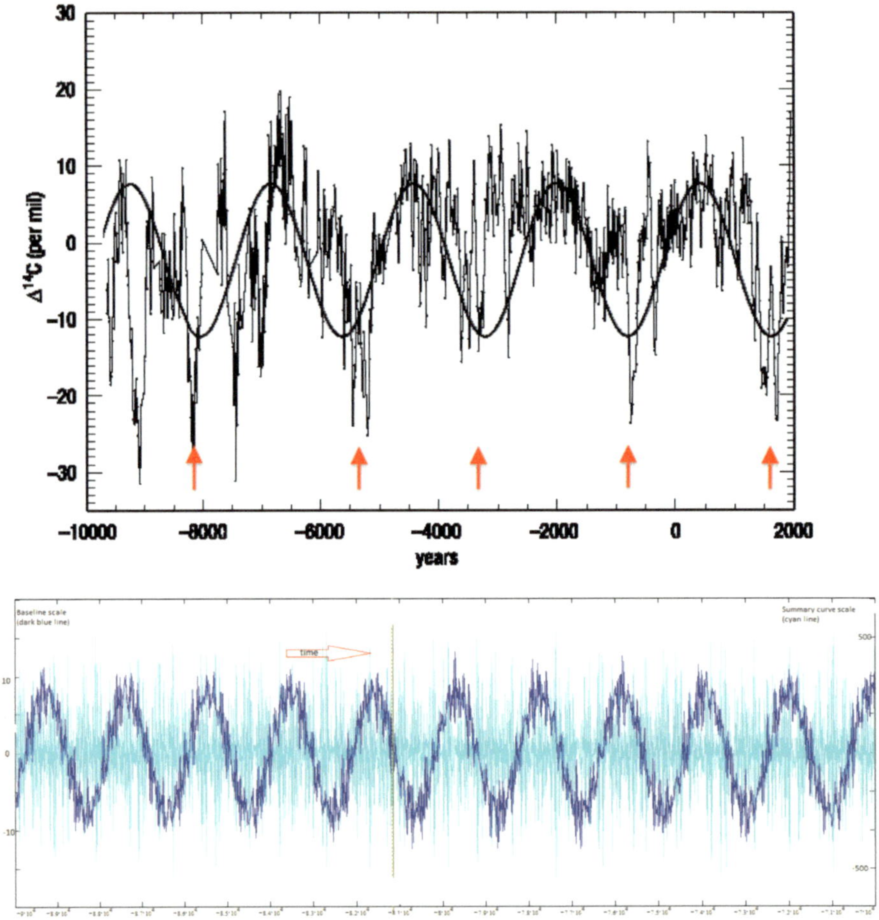


**Figure 3.**
*Top plot: The millennial oscillations with a period (2100–2200) (Hallstatt cycle) of the carbon $^{14}C$ isotope abundances reported in parts per thousand (per mille, %) used for solar irradiance dating in the IntCal09 data from Reimer et al. [39]. This period is similar to that derived from the solar irradiance restored in the past 12,000 years with a wavelet transform by Steinhilber et al. [22] (see their* ***Figure 4****). The positive sign means excess and the negative sign means deficit of abundances. Bottom plot: The oscillations of the baseline (zero line, see for details section 2.2) of solar background magnetic field (left Y axis, arbitrary units, navy line) with a period of about 2000±95 years over-plotted on the oscillations of the reduced summary curve (right Y-axis, arbitrary units, cyan line). Positive magnitudes of the summary and baseline curves represent northern magnetic polarity while the negative ones - southern magnetic polarity. Adopted from Zharkova et al. [19].*

because it was redacted to a single point per year instead of 13 points (for Carrington rotations) originally used [10, 17].

Hence, the baseline magnetic field in **Figure 3** (bottom plot) reveals the very stable oscillations with a period of $T_{base} = 2000 \pm 95$ years [19]. Evidently, these baseline oscillations are normally incorporated into the magnetic field measurements of the summary curve (cyan curve) and, thus, are not detected in the unfiltered observations. The baseline oscillations of magnetic field have a very stable period maintained during the whole duration of simulations of 120 thousand years meaning these oscillations of the baseline magnetic field on a millennial timescale to be induced by a rather stable process either inside or outside the Sun.

The variations of the magnetic field baseline oscillations for the current Hallstatt's cycle are shown in **Figure 4** (top plot, navy line) (from [19]) indicatng that it started at Maunder minimum and is in ascending phase now [20, 42] and the

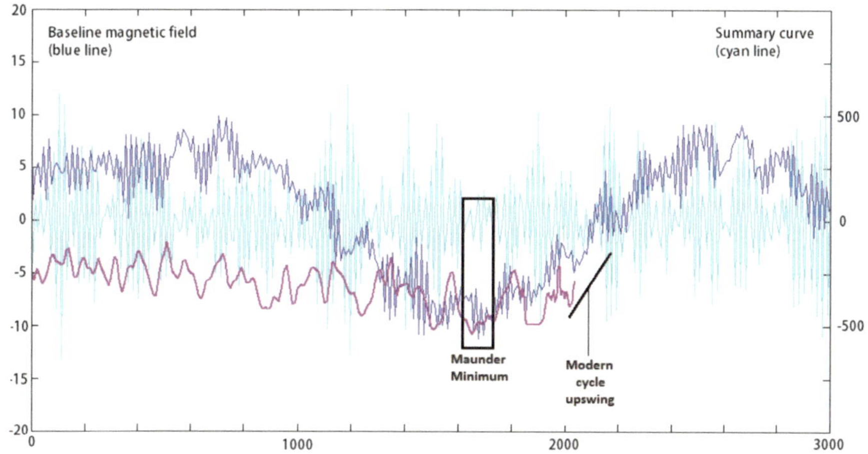


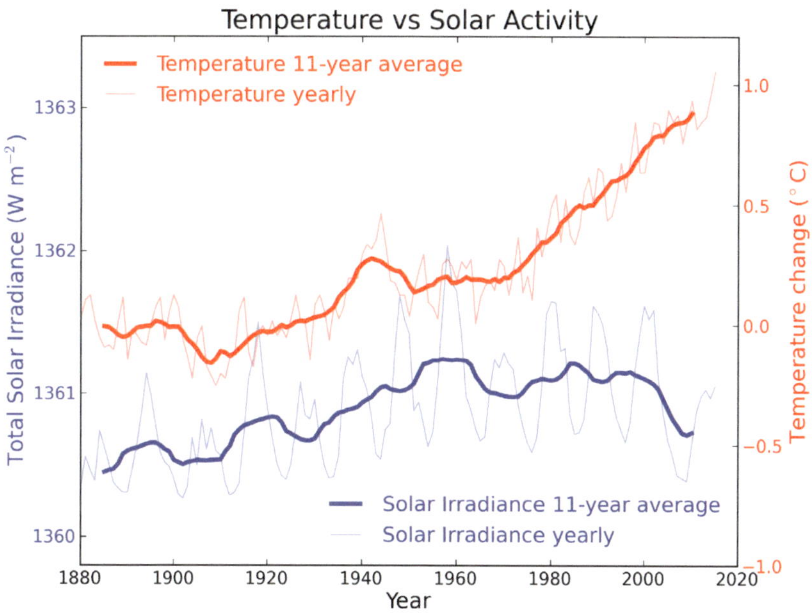


**Figure 4.**
*Top plot: The close-up view of the current cycle of the baseline magnetic field (dark blue curve, arbitrary units, see for details section 2.2) with the minimum occurring during Maunder minimum. The scale of the baseline variations are shown on the left hand side of Y axis, the scale of the summary curve variations - on the right hand side Y-axis. The irradiance curve (magenta line) taken from [20, 39], their* ***Figure 3*** *(top plot), over-plotted on the summary curve of magnetic field (cyan curve) [19]. The irradiance curve had to be reduced in magnitude to avoid full overlapping with the baseline magnetic field curve. The black line defines the slop of the baseline terrestrial temperature from [35]. Adopted from Zharkova et al. [19]. Bottom plot: The variations of terrestrial temperature (red lines) and total solar irradiance (measured in $W/m^2$) (blue lines) during each solar cycle (thin lines) and the one averaged per cycle (thick lines) derived by [45].*

reduced summary curve of magnetic field (cyan line). The irradiance curve was reduced in magnitude by factor 3, in order to distinguish this curve from the baseline oscillations (e.g. Spearman's correlation coefficient between these two curves is about 0.68). After the MM the magnetic baseline curve is growing towards northern polarity, while the solid dark line showing the rate of increase of the baseline terrestrial temperature [35].

From the close-up plot of the current millennial baseline cycle in **Figure 4** (top plot) it becomes evident that from 1600 the baseline magnetic field was shifting towards the northern polarity approaching its maximum in about 2600. This increase of the baseline magnetic field of northern polarity is likely to coincide with the increase of solar irradiance curve [20, 42]. The baseline terrestrial temperature curve is shown increasing by 0.5°C per 100 years [35] and has a slope (shown by the black line in **Figure 4** top plot) close to that of the magnetic field baseline increase (navy line) [19]. At the same time, the variations of the terrestrial temperature versus solar activity shown in **Figure 4** (bottom plot) [45] reveal that in the past few decades the Earth temperature increase goes against the solar activity showing the signs of decrease. This raised some reasonable questions about the cause of the terrestrial temperature increase and led to suggestions of substantial extra-heating of the Earth atmosphere caused by greenhouse gases.

## 3. Millennial variations of the Sun-Earth distance

The observations of solar irradiance and magnetic field baseline oscillations with the period of about 2100–2200 years are believed to be imposed on the Sun by the gravitational effects of Neptune and Saturn causing SIM [30, 46]. In this section we carry out the investigation of Sun-Earth distance variations over a millennial scale in an attempt to establish if they follow or deviate from Kepler's laws.

### 3.1 Ephemeris of the Sun-Earth distances in 600–2600

Let us now explore the daily Sun-Earth (S-E) distances over the two millennia (600–2600) derived from the ephemeris of VASOP87 - Variations Seculaires des Orbites Planetaires [47] *http : //neoprogrammics.com/vsop87/planetary*$_d$*istance*$_t$*ables/*. Note that the VSOP87 data up to 6 digits after the decimal coma coincide with the widely used JPL ephemeris [48].

The daily Sun-Earth distances for every month of the three years for each millennium: M1 (600, 1100, 1600) and M2 (1700, 2020, 2600) are presented in **Figure 5** (January–June) and 6 (July–December) for the millenniums M1 (600–1600) (left column) and millennium for M2 (1600–2600 (right column). The Sun-Earth distances change rather differently over the two millennia M1 and M2 from what one would expect from the elliptic motion of the Earth about the Sun where the perihelion (shortest distance) and aphelion (longest distance) occur on the semi-major axis of the ellipse and the distances are defined by Kepler's third law (see **Figure 15** and Eq. (6) in Appendix A).

However, instead of it, one can observe a significant reduction of the Sun-Earth distances in January–June (**Figure 5**) and their increase in July–December (**Figure 6**). Furthermore, the maximal differences, or differences between the S-E distances at the start and end of each millennium considered presented in **Figure 7**, reveal these maximal differences reaching 0.005 au in April–May (**Figure 5**) in millennium M1 and up to 0.011 au in April–May in millennium M2. These are followed by significant increases of the S-E distances in August–December shown in **Figures 6** and 7. Moreover, the daily double differences, e.g. the differences between the maximal differences of the S-E daily distances in M1 and M2, taken from **Figure 7**, plotted in **Figure 8** demonstrate that in March–June there is a large reduction of the S-E distances in M2 compared to M1.

In order to evaluate if these changes are symmetric, let us present the mean monthly S-E distance variations during each sample year considered plotted in **Figure 9**. This, in fact, reveals that in M1 the increases/decreases of the S-E distances (left plot) are nearly symmetric over each year and centred about the summer solstice in June and winter solstice in December while in the millennium M2 the

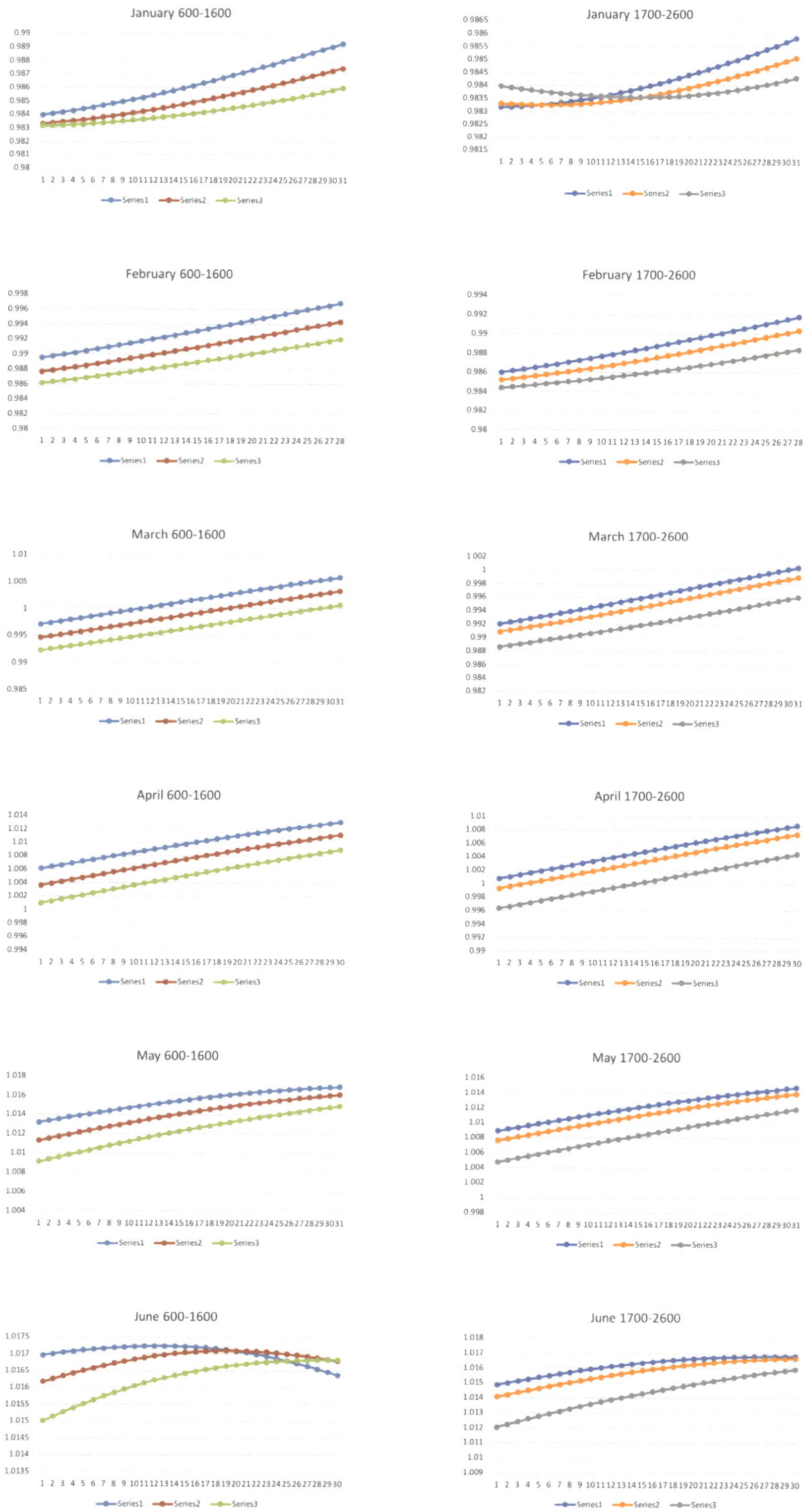


**Figure 5.**
*Variations of the sun-earth distances (in astronomical units, au) versus days of the month (X-axis) in January–June for three sample years in the millennium M1 (600–1600) (left) and M2 (1600–2600) (right). Left column: Blue - year 600, red –1100 and green –1600; right column: Blue - year 1700, red - 2020 and grey - 2600.*

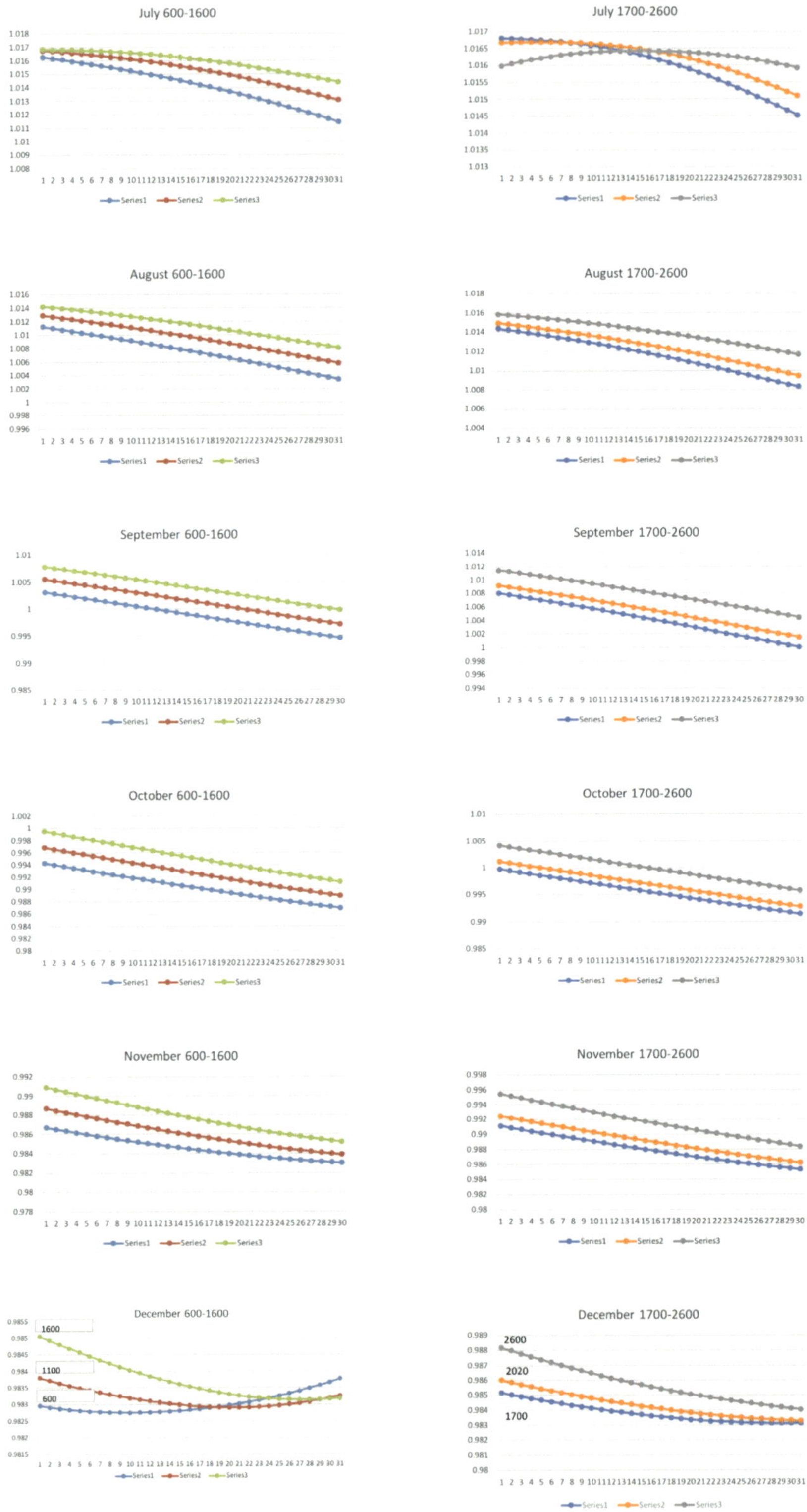


**Figure 6.**
*Variations of the daily sun-earth distances (in astronomical units, au) versus days of the months in July–December of three sample years selected in the millennium M1 (600–1600) (left) and M2 (1600–2600) (right. Left column: Blue - year 600, red –1100 and green –1600; right column: Blue - year 1700, red - 2020 and grey - 2600.*

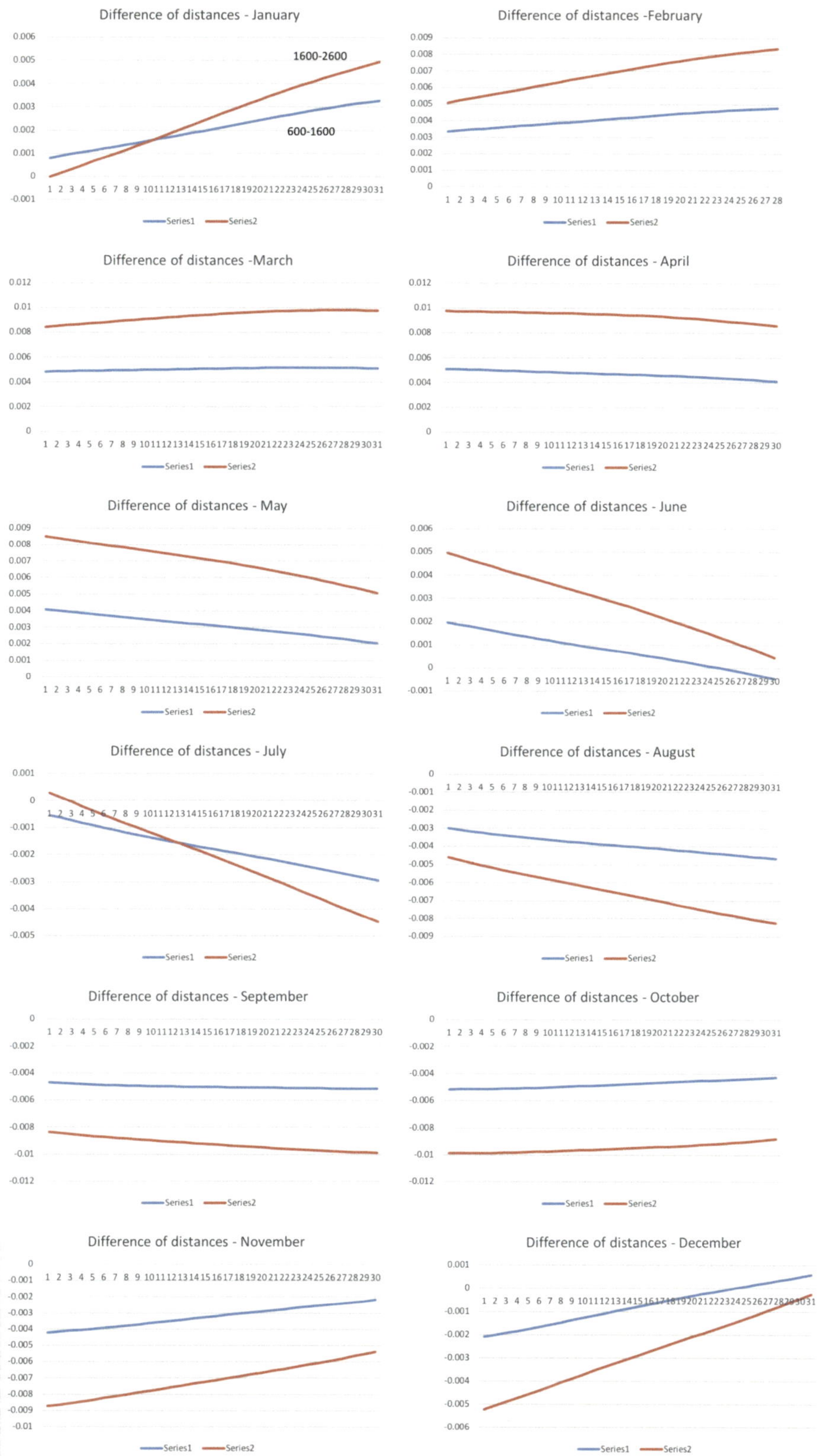


**Figure 7.**
*Maximal daily differences for each months of the sun-earth distances (in astronomical units, au) between the years 600–1600, M1 (blue curves) and 1700–2600, M2 (red curves). X-axis shows days of the months.*

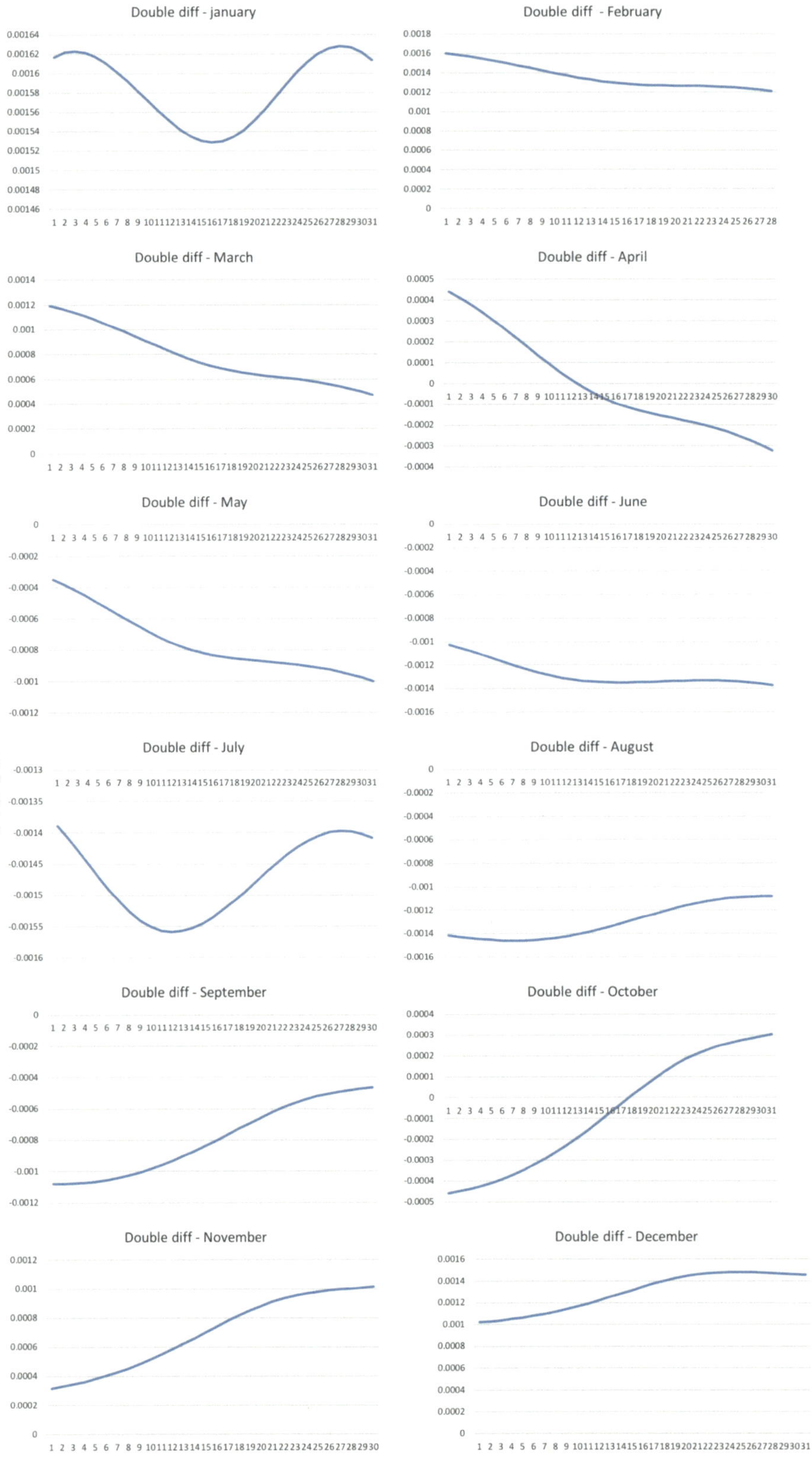


**Figure 8.**
*The differences between the maximal variations of the daily sun-earth distances (in astronomical units, au) in millennium M1 (600–1600) and M2 (1700–2600) shown in* **Figure 7**. *X-axis shows days of the months.*

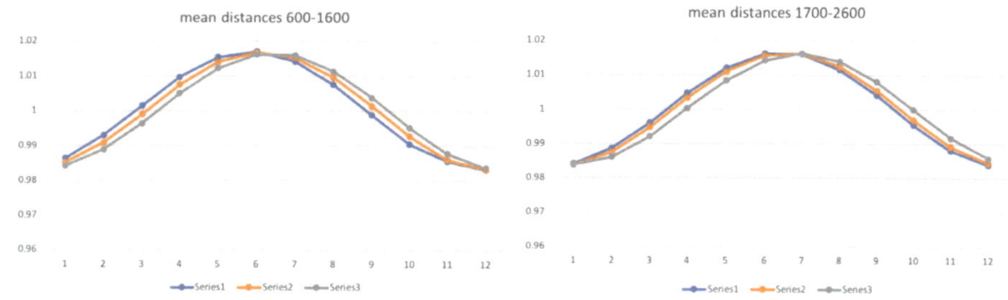


**Figure 9.**
*Variations of the annual sun-earth distances (in astronomical units, au) versus months of the year in the millennia M1 (left) and M2 (right). Left plot: Blue curve - year 600, red curve −1100 and grey curve - 1600; right plot: Blue curve −1700, red curve - 2020 and grey curve - 2600.*

distance curve is skewed (right plot) with the maximal and minimal Sun-Earth distances being noticeably shifted in time towards a mid-July for aphelion and mid-January for perihelion. Namely, in M1 the local perihelion and local aphelion are shifted forward by 5–6 days to 26–27 December and 26–28 June, respectively, from the summer and winter solstices on 21 June and 21 December accepted for elliptic orbit of the Earth revolution about the ellipse focus. While in M2 the local perihelion and aphelion in 2600 are shifted from the elliptic orbit positions for the winter and summer solstices forward by up to 25–26 days (to 15–16 January and 15–16 July, respectively, seen in the right column of **Figures 5** and **6**.

This asymmetry in the changes of the S-E distances in M2 compared to M1 is more clearly demonstrated by the annual variations of the double differences between the S-E distance shown in **Figure 8** after they are averaged for each month and presented over a year in **Figure 10** for each sample years considered. It clearly shows that the shifts in the S-E distances are reduced more in the April–September and increase more in October–February of each year of millennium M2. This means that the input of solar irradiance to the Earth is not evenly distributed over time of the Earth revolution, or over the Earth location on the orbit.

### 3.2 Proposed interpretation of the S-E distance variations

If the Earth revolves about the Sun located in the focus of the ellipse, the Sun-Earth distance has to change depending on the Earth position on the orbit following Kepler's 3rd law (see Eq. 6 in Appendix A). Earth orbit is a stable elliptic orbit with little changes of the major and minor axis, as established in Appendix C with the

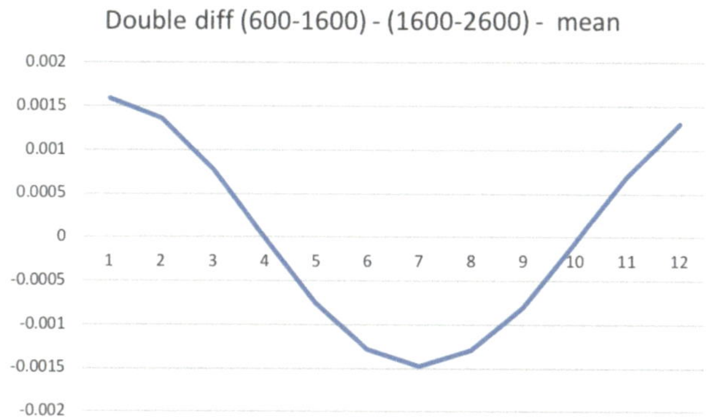


**Figure 10.**
*Annual variations of the differences between the mean maximal monthly differences in the sun-earth distances (in astronomical units, au) in millennium M1 (600–1600) and M2 (1700–2600) taken from* **Figure 8**. *Axis X shows months of a year.*

help of Appendices A and B. However, the S-E distance reductions and growths reported here deviate from the Kepler's third law (Eq. 6 in Appendix A). By comparing the mean-by-time and mean-by-arc S-E distances for an elliptic orbit (see Appendices B) with the expected changes imposed by the calculated shifts of aphelion and perihelion [49] shown in Appendix C, it is evident that the real S-E distances derived from the ephemeris are different from Kepler's 3rd law (see Eq. 6 in Appendix A).

This can can only happen if these S-E ephemeris reflect the additional motion: the revolution of the Sun about the barycentre, which is induced by the action of large planets of the solar system. The similar effect is observed in the stars, which have planetary systems, leading to a wobbling star effect that is used to trace possible exoplanets [50, 51]. The shift of S-E distances reported above should be caused by the increasing shift of the Sun's location from the focus of the ellipse, where it is supposed to reside, according to Kepler's laws, towards the spring equinox of the Earth orbit. This shift of the position of the Sun with respect to the barycentre has been recognised as the solar inertial motion - SIM [25, 27, 30]. The resulting S-E distances are defined by the superposition of these two motion: Earth revolution and SIM.

In fact, the variations of the S-E distances during the two millennia are likely to be affected by the gravitational effects of Jupiter, Saturn, Uranus and Neptune on the Sun's inertial motion [30] revealing the oscillations of the planet orbits with a period of 8.5 thousand years (see Fig. 1 in [30], affecting SIM. From the whole period of 8.5 thousand years reported in the paper the semi-period with maximum of 4.2–4.3 thousand years with the ascending part of 2.1 thousand years are similar to the period of decreasing S-E distances reported in section 3.1 for 600–2600. Also the reported S-E distances reveal the noticeable shifts of the aphelion and perihelion from the major axis of the ellipse that coincides also with the oscillations of magnetic field baseline [19, 52] and solar irradiance [22]. It seems that in the two millennia 600–2600 the large planets continuously shifted the Sun from its focus towards the spring equinox as detected from the S-E ephemeris in **Figures 5** and **6**.

Therefore, it can be noted that owing to SIM, the shortest and longest Sun-Earth distances (perihelion and aphelion) in the elliptic orbit of the Earth are shifted to the local aphelion and perihelion, which are located on the shorter axis of the ellipse than the major axis. This line has an angle $\phi$ to the semi-major axis roughly defined by the formula for tan $\phi$:

$$\tan\phi = \frac{2d_s}{f}, \tag{1}$$

where $f$ is a distance between the foci of the ellipse and $d_s$ is the shift along the semi-minor axis $b$ of the Sun from the focus of the ellipse. Naturally, by the definition of an ellipse, this line is shorter than the semi-major axis $a$ of the Earth elliptic orbit, which is the longest axis in the ellipse.

Furthermore, the calculations of the double differences between the maximal distance shifts occurred in millennia M1 and M2 (M1-M2) for daily data shown in **Figure 8** and their annual variations shown in **Figure 10** reveal that the double differences become negative in April and remain such until the end of October. This means that in M2 (1600–2600) the S-E distance decreases in April–July and its increases in July–December are much larger that in M1 (600–1600). This also indicates that in M2 the Sun becomes closer and closer to the Earth during April–October before the Earth revolution will make the S-E distance increases in November–February, since these increases are larger than expected from Kepler's third law. This, in turn, can lead to a significant solar radiation input to the Earth in millennium M2, which needs to be processed by the Earth atmosphere and ocean that will be discussed below in section 4.

These long-term SIM effects can explain the reported above significant S-E distance decreases in January–June and the similar increases in July–December during the both millennia. The magnitudes of the S-E distance oscillations are smaller for M1 (up to 0.005 au) and twice larger for M2 (0.011 au) shown in **Figures 5** and **6**, and specifically in **Figure 7** producing daily differences in the S-E distances for each month of the years for the two millennia considered. In the next two millennia this trend is expected to return back to the level in 600 and then in the next 4.2 thousand years to change to the opposite one, e.g. producing the Sun shift to the autumn equinox and the shifts of the local perihelion and aphelion for considered years towards early December and June, respectively, following the calculations by [30].

Interestingly, the annual variations of the S-E distances shown in **Figure 9** can explain the oscillations of the baseline solar magnetic field (Hallstatt's cycle) shown by dark blue lines in **Figure 3** and in **Figure 4** (top plot) [19] by the oscillation of the Earth aphelion and perihelion from the major axis. In M1 the Sun's location is closer to the ellipse focus of the Earth orbit resulting in a smaller magnitude of the Sun's shift in the direction of the minor axis that leads to the minimum of the baseline magnetic field of northern polarity, shown by the dark blue line in **Figure 3** (bottom plot) [19, 52]. While in M2 the Sun shifts much further from the focus towards the spring equinox position of the Earth orbit, so that there is a shift of the longest S-E distance (local aphelion) from 21 June (when the aphelion on the major axis of ellipse is approached) to 16 July when the aphelion is shifted from the major axis to the line of the ellipse connecting the ellipse centre and displacement of the Sun from the ellipse focus and directed under the angle $\phi$ (see Eq. (1)) to the major axis.

Hence, in 1600–2600 the Earth will be turning closer to the Sun for up to 25 additional days after the summer solstice, while turned towards the Sun with its Northern hemisphere, before it approaches the local aphelion. This is likely to cause a small rise to the baseline magnetic field of northern polarity as shown in **Figure 3** (dark blue line) [19]. And given the periodic variations of the gravitational effects of four large planets described by [30], one can expect the similar periodic variations of the baseline magnetic field linked to the positions of the local aphelion and perihelion for a given epoch. Therefore, this confirms the hint expressed earlier [19, 52] that the baseline magnetic field oscillations derived there purely from the magnetic field observations are, indeed, caused by the gravitational effects of large planets on the Sun, or by solar inertial motion.

## 4. Millennial oscillations of solar irradiance with the Sun-Earth distances

### 4.1 Method of inverse squares

Following the variations of the S–E distances discussed in section 3, let us evaluate the variations of total solar irradiance (TSI) imposed by a change of these S-E distances in the millennia M1 and M2 using the method of inverse squares. A magnitude of the total solar irradiance $S$ variations at the solar-Earth distance $d$ by considering the Sun as a point body emitting radiation with an intensity $I_{\odot}$ [53]:

$$S = \frac{I_{\odot}}{d^2}. \tag{2}$$

Hence, the solar irradiance $S$ can vary either because of the variations of intensity $I$ of solar radiation at the Sun itself or because of the variations of a distance $d$ between the Sun and Earth. The variations of the solar intensity $I$ is caused by the variations of solar activity induced by the electro-magnetic dynamo action in the solar interior.

If the intensity $I_{\odot}$ of radiation on the Sun is considered to be constant at a given time ($I_{\odot}$=const), then the solar irradiance $S$ can also change because of a variation of the Sun-Earth distance caused by the Earth orbital motion itself leading to the terrestrial seasons and by solar inertial motion whose effects are not yet fully investigated. In any case, by knowing the ephemeris of the S-E distances and using Eq. (2) above for calculating solar irradiance at two different distances $d_1$ and $d_2$, one can find the relationship between the solar irradiance, $S_1$ and $S_2$ at these distances, which follows the inverse square law [53]:

$$S_1 \cdot d_1^2 = S_2 \cdot d_2^2. \quad (3)$$

Therefore, if at a distance $d_1$ the average solar irradiance is 1366 $W/m^2$ [22, 31] then if the distance is changed to $d_2$, the solar irradiance $S$ should also change following the Eq. (3). For example, if the distance $d_2$ between the Earth and Sun was to be decreased by 0.016 au (as shown in section 3 for two millennia 600–2600) so that the initial irradiance of 1366 $W/m^2$ divided by the square of the new distance results in the irradiance of 1411 $W/m^2$. The difference in the irradiance is 1411–1366 = 45 $W/m^2$, that is 3.3% that is exactly the magnitude mentioned in the first paragraph of the last section of paper [19].

In section below the solar irradiance is explored in more details for the two millennia from 600 to 2600 AD for the S-E distances presented in section 3.

### 4.2 Orbital variations of solar irradiance in the millennia 600–2600

As established in section 3, the Sun-Earth distances are changing accordingly to the ellipse curve as Kepler's 3rd law assigns. Instead, these distances are defined by the two motions: the Earth and Sun about the barycentre of the solar system with the latter caused by the gravitational effects of large planets of the solar system, or solar inertial motion (SIM). Therefore, the daily variations of solar irradiance over a year will be affected by the combination of the Earth revolution on its orbit and the Sun's revolution about the barycentre.

By using Eq. (3) let us calculate the solar irradiance at any day of a year during the two millennia M1 and M2. For the TSI normalisation the magnitude of $S_1$ = 1366 $W/m^2$ [31] can be used for the longest distance in June 1700. Then the daily TSI magnitudes for every month of a year for three years for each millennium: M1 (600, 1100, 1600) and M2 (1700, 2020, 2600) are presented in **Figure 11** (for January–June), and 12 (for July–December) with their annual variations compared in **Figure 13**. The small differences ($\leq$ 0.001 au) between the S-E distances of 1600 and 1700 are considered when calculating the total solar irradiance for M2. The overall variations of the sum of the TSI deposited to Earth in each year are presented in **Figure 14** calculated for: (a) the mean TSI magnitudes averaged for every month, e.g. by adding the TSI magnitudes for 12 months (left plot) and (b) the daily TSI magnitudes, e.g. by adding the TSI magnitudes for all days in each year considered.

In M1 the increase of solar irradiance during the months January–June is nearly balanced by its decrease from July to December (**Figures 11** and **12**, left column) while in M2 the solar irradiance is noticeably higher in February–July when the

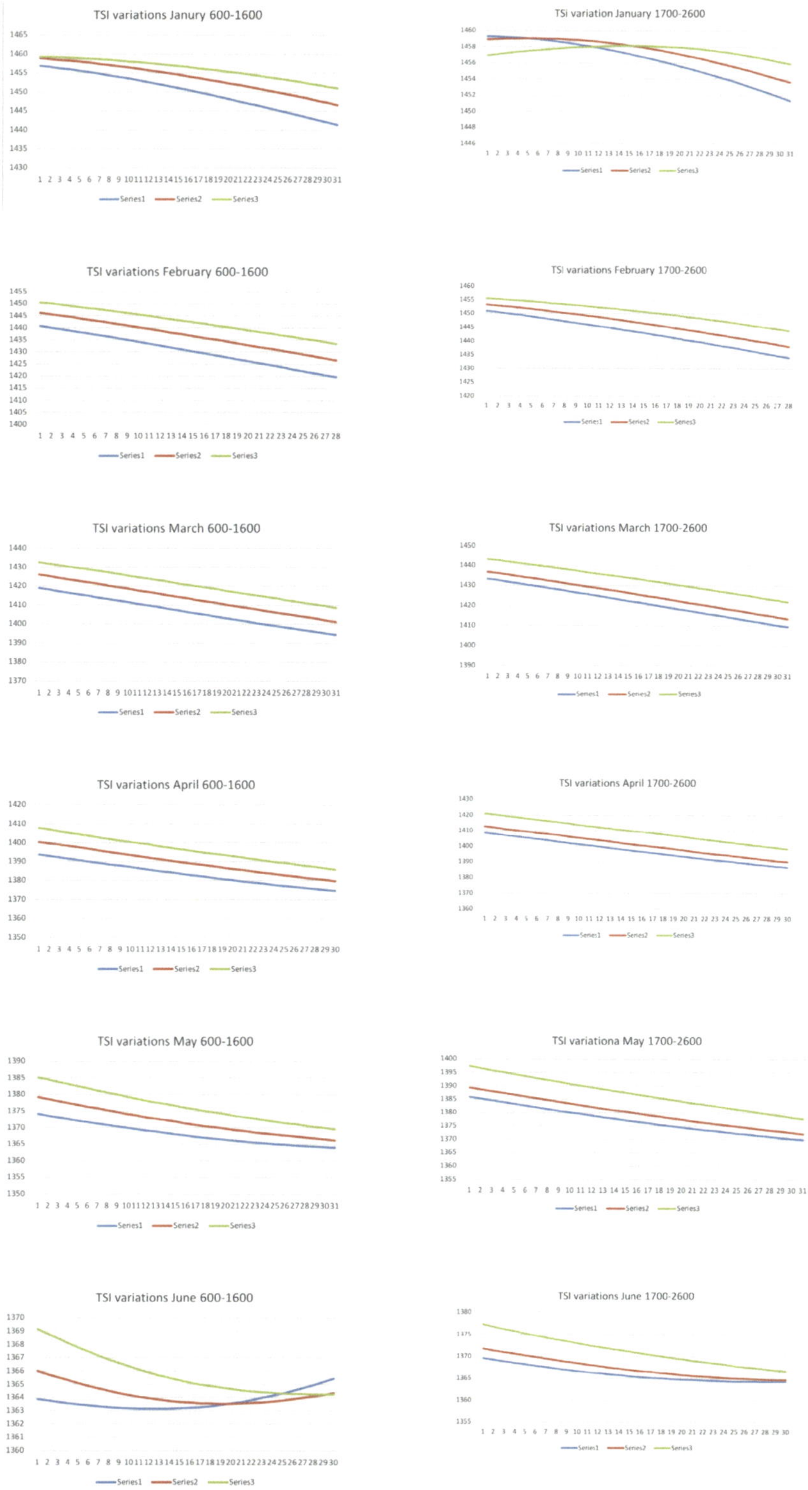


**Figure 11.**
*Variations of the daily solar irradiance in (W/m²) in January–June for three sample years selected in the millennia M1 (600–1600) (left) and M2 (1600–2600) (right). Left column: Blue - year 600, red –1100 and green –1600; right column: Blue - 1700, red - 2020 and grey - 2600. X-axis shows days of the months.*

**Figure 12.**
*Variations of the daily solar irradiance ($W/m^2$) in July–December of three sample years in the millennium M1 (600–1600) (left) and M2 (1600–2600) (right). Left column: Blue - 600, red –1100 and green –1600; right column: Blue - 1700, red - 2020 and grey - 2600. X-axis shows days of the months.*

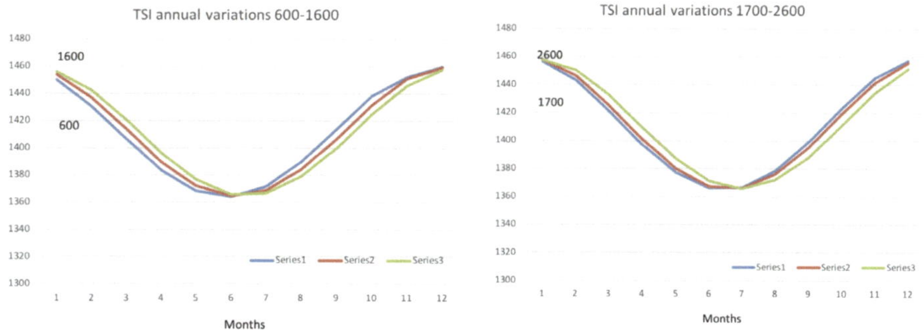


**Figure 13.**
*The annual variations of TSI magnitudes ($W/m^2$) in millennia M1 (600–1600) (left) and iM2 (1700–2600) (right). Axis X shows months of a year.*

atmosphere is heated by the Sun than in July–December when the atmospheric cooling occurs (see **Figures 11** and **12**, right column). This is more evident in the annual variations of the monthly averaged TSI magnitudes (**Figure 13**) revealing a steady increase of the solar irradiance input in millennia M1 and, especially, in M2 during spring-summers and decrease during autumn-winters in the Northern hemisphere in each century caused by the variations of S-E distances shown in **Figure 9** discussed above.

Because of a reduction of the S–E distances in the first half of a year caused by SIM, the TSI deposition from years 1700 to 2600 is increased by about 11 $W/m^2$ (0.95%) in February–March (and decreased by the close amount in August–September), by 15 $W/m^2$ (1.2%) in April–May (decreased in October–November) and by 7–8 $W/m^2$ (0.5%) in June–July (decreased in December–January) (see **Figures 11** and **12**). These TSI variations can naturally explain a wide variety of the measured TSI magnitudes in the earlier space observations of 1370 $W/m^2$ (Shirley et al. 1990), 1971 $W/m^2$ (Wolff & Hickey 1987), or 1972 $W/m^2$ (Lee III et al. 1995) if they are measured during May–June or July–August. The numbers of TSI variations during a first half of a year can be added to produce more than 2.7% of solar irradiance increase in M2 because of the S-E distance decrease by SIM that is comparable with the estimations up to 3.5% hinted in the retracted paper [19]. *This amount* of the extra solar radiation input into the terrestrial atmosphere and ocean *has not been yet considered in the current climate models.*

The variations of solar irradiance averaged for every month in a year are plotted for both millennia in **Figure 13**, showing that the minimum of the mean solar irradiance is shifting in M2 towards 15 July, thus, securing the extra heating of Northern atmospheres in the summer months of second half of June and half of July in this millennium M2. These shifts of the largest S-E distances aphelion from 21 June to 15 July (local aphelion) in M2 can also explain why the baseline solar magnetic field is an ascending phase of Hallstatt's current cycle, with a maximum of the northern polarity at 2600 before the longest distance becomes shifting back to June in the next few millennia. As shown in [19] (see **Figure 3**) there have been about 60 super-grand (Hallstatt's) cycles over the past 120,000 year. This means such the millennial changes of the TSI on Earth are regular patterns, which will continue to appear in the current Hallstatt's cycle shown in **Figure 4**, top plot [19].

Based on the location of Earth on its orbit, these solar irradiance inputs has to be divided between the hemispheres depending on which one of them is turned towards the Sun. This means that, because of the Earth axis tilt of 23.5° from the vertical to the ecliptics, in the millennium M1 (600–1600) the input of solar radiation in the Northern hemisphere was slowly increasing from January until 21 June not only because of the elliptic Earth orbit but also because of the Sun's shift from the focus of this ellipse in the minor axis direction towards the spring equinox and become reducing from 21 June through the whole July. While in M2 (1600–2600) during the months June – July the input of solar irradiance to the Northern hemisphere will be higher than in the elliptic orbit. This means that in M2 the increase of the solar input in February–July must be ahead of its decrease in August–January. This would happen also because, according to Kepler's second law, the Earth moves slower at the parts of the orbit in June–July–August than in December–January, thus, passing quicker through the positions with a reduced radiation in December than with the increased one in June–July.

### 4.3 Imbalance of the TSI depositions in the two millennia

Since there is a shift of the minimum point of the TSI annual variations (**Figure 13**) from 21 June (M1) to 15 July (end of M2), this indicates a possible imbalance between the annual TSI input and output in M2 (1600–2600). From the daily magnitudes of TSI shown in **Figures 11** and **12**, it is possible to count the total annual amount of TSI emitted by the Sun towards the Earth in each year of the both millennia. If this amount does not change from year to year, then TSI is, indeed, the same for each year for both centuries, as currently assumed.

However, the real annual magnitudes of TSI deposited to the Earth during the two millennia are shown in **Figure 14** calculated for the two cases: (a) added together the averaged monthly TSI magnitudes (left plot) calculated for the S-E distances shown in **Figure 13** when only 12 magnitudes per year (for 12 months);

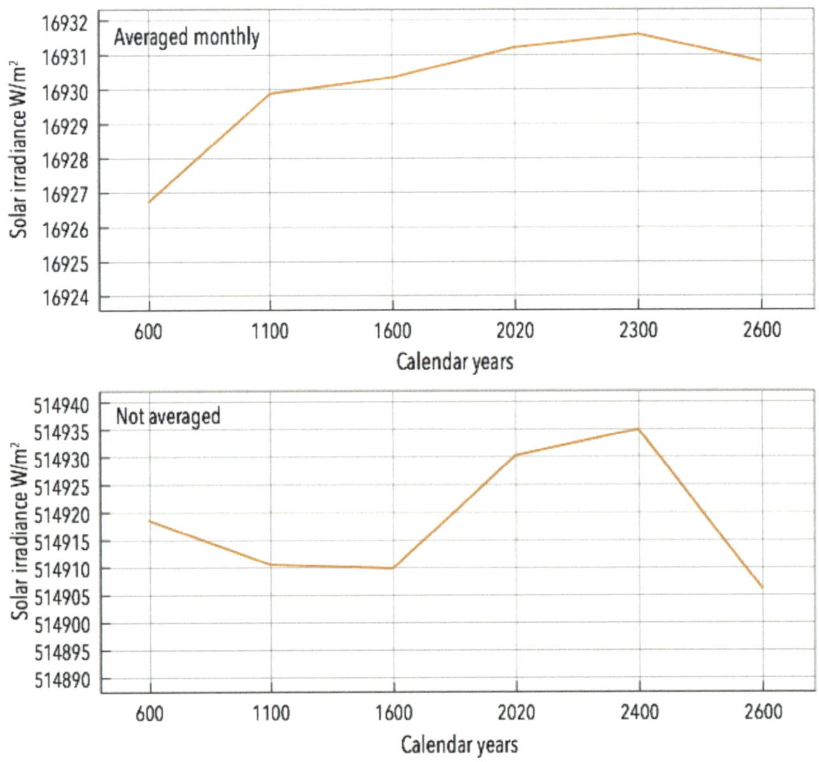


**Figure 14.**
*The total annual TSI variations ($W/m^2$) in the millennia M1 and M2 derived by summation of the mean monthly (top) and daily TSI magnitudes (bottom). Axis X shows the years of the millennia.*

(b) added together the daily TSI magnitudes (right plot) taken from **Figures 11** and **12** associated with the daily magnitudes of TSI (for 366 days for the leap years used).

These two plots clearly demonstrate that the monthly TSI variations (case a) show the increase of TSI by about 1–1.3 $W/m^2$ in 2020 compared to 1700 (**Figure 14**, left plot). This TSI increase found from the S-E distance ephemeris is close to the magnitude of 1–1.5 $W/m^2$ reported from the current TSI observations [34]. However, the annual TSI magnitudes, calculated from the daily S-E distances (case b) reveal a much larger annual increase of the total solar irradiance by about 20–25 $W/m^2$ ( > 1.8%) in M2 (by 2500) than in millennium M1 (**Figure 14**, right plot). This analysis gives the indication of the averaged TSI increase in M2 could be 2.5–2.8 $W/m^2$ per century, or (0.18–0.20)%, comparing to the TSI in 1700. This is the very important hidden solar irradiance input in millennium M2 (1600–2600) caused by the SIM effects, which was significantly underestimated if only the averaged monthly TSI magnitudes are used (**Figure 14**, compare left and right plots). The essential issue is how much of this extra solar radiation is distributed between the hemispheres owing to the Earth tilt, its position on the orbit or the level of exposure to solar radiation [42, 49].

Our study of the S-E distance variations shows that at the start of any year, in January, the Earth is turned to the Sun with its southern hemisphere, meaning that any decrease and increase of solar radiation during this time is mostly absorbed by the parts in Southern hemisphere. When the Earth's orbiting approaching March, the distribution of solar irradiance between the hemispheres becomes nearly even, while in April–June the main part of the solar radiation input is shifted towards the Northern hemisphere, having its maximum theoretically (by Kepler's law) on 21 June, while in reality, shifted to 5 July in 2020 and to 16 July in 2500. Hence, in M2 the Northern hemisphere should get the extra solar radiation not only in the first six months of a year but also in the 25 days from 21 June by approaching the local aphelion on 16 July, which is not compensated later by its expected cooling because of a shift of the local perihelion to 16 January.

By comparing the mean-by-time and mean-by-arc S-E distances for an elliptic orbit (see Appendices B and C) based on the calculated shifts of aphelion and perihelion [49] with the real S-E distances derived from the ephemeris one can conclude that the ephemeris of the S-E distances have to reflect the Sun shifts in SIM, in addition to the Earth revolution about the ellipse focus. Therefore, the solar radiation deposition in the millennium M2 is expected to be essentially higher than in in millennium M1 and different from the standard seasonal changes because of the uneven shifts of Sun-Earth distances on the orbit owing to SIM. This extra TSI amount caused by SIM (from the variations of a distance $d$ in formula (2) will undoubtedly add to the magnitude of solar irradiance coming from the solar activity itself (or the parameter $I_{\odot}$ in formula (2)) shown in **Figure 4** (bottom plot, blue lines) leading to the overall solar irradiance increase that, in turn, can account for a large amount of the terrestrial temperature increase shown by the red curves in **Figure 4** (bottom plot). This extra solar forcing caused by SIM needs to be taken into account in any climate models.

## 5. TSI variations and terrestrial temperature

Let us try to evaluate how these variations of solar irradiance can affect terrestrial temperature from the general similarity approach. The TSI variations caused by the solar activity in normal cycles of 11 years and during grand solar minima (similar to Maunder Minimum) can be described as follows.

1. Solar irradiance $S$ variations at Earth owing to 11 year cycle is about 0.1% of the average magnitude of TSI $S$ (1366 $Wm^{-2}$ accepted in this study) increasing by 1.4 $Wm^{-2}$ during maxima and decreasing during minima [33, 54]. The terrestrial temperature variations during 11 year cycle are negligible.

2. Solar irradiance $S$ variations at Earth owing to GSM is about 2.5–3 $W/m^2$, or 0.22 % of $S$ [31, 37, 55] as shown in **Figure 2** (left plot). These estimations are also supported by conclusions by other authors [55, 56] showing sometimes up to 0.4% contributions of active regions into the solar radiance intensity $I_{\odot}$.

The terrestrial temperature curve presented in **Figure 2** (right plot) shown a reduction during MM of the average terrestrial temperature by about 1°.0C [36, 57, 58], e.g. the decrease of TSI by 0.11% secures a decrease of the terrestrial temperature by approximately 0°.5C. Let us use this simple estimation until we carry out more precise model simulations.

### 5.1 Expected effects of the TSI increase by SIM on the terrestrial temperature

The terrestrial temperature is found increasing since Maunder minimum as shown in **Figure 4**, bottom plot derived by Akasofu [35] that is close to the plot presented in the NASA and IPCC report *https : //www.ipcc.ch/sr15/*. At the same time the solar activity of 11 years, and thus, solar irradiance caused by it in the past four solar cycles was decreasing.

Now we established that there is an additional effect leading to the increase of solar irradiance in the millennium M2 (1600–2600) because of the changing Sun-Earth distances imposed by the solar inertial motion (SIM) owing to gravitational effects from Jupiter, Saturn, Neptune and Uranus. The overall increase of solar irradiance for M2 is shown in **Figure 14** to reach about 20–25 $W/m^2$ for the whole planet, which can be assumed to split evenly to each hemisphere with 10–12 $W/m^2$.

Although, the conversion of this extra solar radiation into the terrestrial temperature is a complex process involving exchanges between the deposited solar radiation to different hemispheres, ocean and atmospheric radiative transfer [59]. In fact, using line-by-line radiation transfer (LBL-RT) calculations under different cloudiness conditions, ground temperatures, and humidity models for radiative transfer of UV solar radiation by atmospheric molecules including $CO_2$, Hardy [59] has shown that even a smaller increase of solar radiation by 5 $W/m^2$ leads to a noticeable (60%) increase of the terrestrial temperature defined by the Sun and only 40% defined by the $CO_2$ emission. The further increase of solar irradiance owing to the millennial TSI misbalance derived here in section 4.3 from the ephemeris of the Sun-Earth distances would definitely lead to a further contribution (possibly, above 80%) of the Sun's radiation into the observed terrestrial temperature growth.

Although, in the current study we do not carry out radiative transfer simulations, and thus, can only roughly estimate possible variations of the average terrestrial temperature using the observed curves similar to those measured [35, 45] (**Figure 4**, bottom plot). The baseline temperature was shown to increase, or to recover from 'little ice age' after Maunder minimum, in the past three centuries (black straight line in [19, 35]). Since the TSI increase by up to 25 $W/m^2$ for two hemispheres, or 12.5 $W/m^2$ per hemisphere is expected until, at last, 2500, then using the link between the solar irradiance and terrestrial temperature derived from **Figure 2**, the increase in the baseline terrestrial temperature from 1700 can be expected by about 4.0°C in 2500, or by 2.0°C in 2100 and by 1.5° C in 2020.

However, these are rather rough estimations. Further investigation of the level of conversion of solar radiation into the atmospheric heating and radiation of terrestrial atmosphere using radiative transfer simulations are required. This can provide more accurate numbers for the terrestrial temperature variations caused by the increase of solar irradiance owing to solar activity and SIM, in general, and their fluctuations in the hemispheres, in particular.

Note, this proposed prediction of the baseline temperature variations does not explain further temperature fluctuations above the baseline temperature which can well be caused by either anthropogenic or other terrestrial activities not considered on this paper.

### 5.2 Effects of upcoming grand solar minimum (2020–2053)

Although in the next 33 years the Sun is entering a period of the reduced solar activity, the modern grand solar minimum, which can be called a 'mini ice age', similar to Maunder Minimum. The GSMs are caused by significantly reduced solar magnetic field imposed by the disruptive interference of two magnetic waves generated by the double dynamo in the solar interior [10]. The first modern GSM1 occurs in 2020–2053 [10, 60] and the second modern GSM2 will happen in 2370–2415 [10, 60].

Because the solar irradiance and terrestrial temperature already increased since the MM owing to the SIM effects discussed in section 5.1, the terrestrial temperature during the first modern GSM1 is expected to drop by about 1.0C to become just (1.5–1.0=) 0.5° C higher than that in 1700.

The temperature decrease during the second modern GSM (2375–2415) can be estimated calculated as follows. The current temperature increase in 2020 is by 1.5°C, which should increase by 2375 by another 1.5° C (=3 x 0.5C [35]) giving the total increase since 1700 by 3.0° C. The temperature decrease caused by a reduction of solar magnetic field and solar activity during the GSM2 would lead to a reduction of temperature by about 1.0° C. This will produce the total temperature during the GSM2 of (3.0–1.0=) 2.0°C higher than in 1700. After each of the modern GSMs, solar activity is expected to return to normal 11 year cycles as shown in **Figure 1** [10].

## 6. Conclusions

In this chapter the investigation of Sun-Earth distances from the ephemeris by VSOP87 [47] and JPL ephemeris [48] is presented. The Sun is found shifting in millennia M1 and M2 along the direction of the minor axis towards the spring equinox that leads to a significant reduction of S-E distances in January–June by about 0.005 au in M1 and up to 0.011 au in M2, which are followed by the asymmetric increases in the second half of the year (July–December). However, the S-E distance increases and decreases are not identical as expected from elliptic orbit.

These S-E distances are found affected not only by the Earth revolution about the focus of the ellipse, but also by the Sun's motion about the barycentre caused by the gravitational effects of other planets (Jupiter, Saturn, Neptune and Uranus), or solar inertial motion (SIM). This shift of the position of the Sun with respect to the barycentre has been recognised as the solar inertial motion - SIM [25, 27, 30]. The resulting S-E distances are defined by the superposition of these two motion: Earth revolution and SIM. The similar inertial motion effects are often observed in other stars, which have planetary systems, leading to the wobbling star effect that is used to trace possible exoplanets [50, 51].

The S-E distance shifts are found to lead to a migration of the Earth's aphelion and perihelion to their classic position on the major axis of the ellipse to occur on 21 June and 21 December, respectively, appropriate for the ideal elliptic Earth revolution. For example, the aphelion is shifted: in 1600 to 28 June, in 2020 to 5 July and in 2060 to 16 July, while and the perihelion migrates from 21 December to 28 December in 1600, 5 January in 2020 and to 16 January in 2600. The shifts of the S-E distances lead to the shifts of the Earth aphelion and perihelion from the major ellipse axis to the intermediate (shorter) axis, which passes through the SIM position of the Sun for the year and the ellipse orbit centre. Therefore, these shifts define the skewness of Sun-Earth distances along the Earth orbit towards the real position of the Sun, because it is moved outside the focus owing to the orbital perturbations of the Sun's motion about the barycentre caused by the gravitational forces of the four large planets.

These shifts of Sun-Earth distances lead to the changes in the total solar irradiance reaching the Earth atmosphere and baseline magnetic field measured from the Earth. Because of this reduction of the S–E distances caused by SIM, the TSI at the Earth is shown to increase from 1700 to 2600 by about 11 $W/m^2$ (0.95%) in February–March (and decreased in August–September), by 15–18 $W/m^2$ (1.2%) in April–May (and decrease in October–November) and by 7–8 $W/m^2$ (0.5%) in June–July (and decrease in December–January). While the shift of the maximal distance (aphelion) from regular 21 June date in 1600 to mid-July in 2600 can naturally explain the skewness of the baseline magnetic field towards the Northern polarity in 2600 and the minimum of the baseline magnetic field in 1600, by its skewness towards Southern polarity as it was reported before [19, 52].

It is also shown that since 1600 to 2020 there was an increase of the annual TSI magnitude by about 1.3 $W/m^2$ derived from the mean monthly S-E distances, which is close to the magnitude of 1–1.5 $W/m^2$ reported for the similar period from the current TSI observations [34]. However, the annual TSI magnitudes, calculated from the daily S-E distances reveal a much larger annual increase of the total solar irradiance by about 20–25 $W/m^2$ by 2500 in M2 compared to millennium M1. This means there is an excess of solar radiation input into the terrestrial atmosphere in millennium M2 not accounted for by any other consideration that has to be considered for the solar forcing. This additional solar input should have different redistribution between Northern and Southern hemispheres, in addition to normal variations of the Earth position on elliptic orbit [49] linked to their exposure time to the solar input not discussed in the current paper.

However, in 2020 the Sun has entered the period of a reduced solar activity: the Grand Solar Minimum (2020–2053). The orbital variations of solar irradiance will be combined with the variations of solar activity, or solar magnetic field, imposed by the variations of solar dynamo [1, 10]. The decrease of solar irradiance during this GSM is expected to be about 3 $W/m^2$, or 0.22%. Therefore, the reduction of solar irradiance caused by the GSM effect will work in opposition to the increase of solar irradiance caused by the orbital SIM effects in the current Hallstatt's cycle.

The baseline temperature (not including any terrestrial effects) is shown increased by 2020 by 1.5°C since 1700 because of SIM effects. Because of the modern GSM1 the terrestrial temperature is expected to be lowered by 1.0°C giving the resulting temperature of 0.5°C higher than it was in 1700. After 2053, the solar irradiance and the baseline terrestrial temperature is expected to return to the pre-GSM level. Then the irradiance and temperature will continue increasing because of the SIM effects combined with radiative transfer of solar radiation in the terrestrial atmosphere. This means the terrestrial temperature will continue increasing up to 3.0°C by 2375 when the second modern GSM2 will occur (2375–2415). During GSM2 the temperature is to be reduced again by 1.0°C to reaching the magnitudes of 2.0°C higher than it was in 1700.

Therefore, our analysis with the new proxy of solar activity (SBMF) has opened new perspectives for reliable prediction of solar activity on short, medium and long-terms. This approach has allowed us to link the solar magnetic field variations to the variations of solar irradiance, which are associated with both the inner solar processes and the orbital effects on the Sun-Earth distances. The fundamental oscillations of solar irradiance, in turn, can be linked to the oscillations of the baseline terrestrial temperature, independent of any terrestrial processes of radiative transfer and heating. Although, other terrestrial and anthropogenic effects can lead to the fluctuations of this temperature but their study was outside the scope of the current chapter.

## Acknowledgements

V. Zharkova wishes to express her deepest gratitude to the funding by the public supporters raised through 'Fund-me' sites. Their support inspired the author to undertake the investigation of the ephemeris of the Sun-Earth distances and relevant variations of the solar irradiance associated with the changes of the Sun-Earth distances induced by orbital effects. The author also wishes to thank the Paris Observatory, France and JPL ephemeris websites, JPL, Pasadena, US for providing the ephemeris of the Sun-Earth distances for a few millennia. The work was partially supported by the US Airforce grant PRJ02156.

## Appendix

### A. Basics of planetary orbits.

In order to investigate the orbital effects on the distance between the Sun and Earth and the resulting variations of solar irradiance imposed by these variations, let us first remind the basic laws governing the planet revolution about a central star, the Sun. It is suggested that the planets evolution about the central star (Sun) is defined by Kepler's three laws [61]:

1. Planets move in elliptical orbits around the central star (Sun), which is located in one of the foci of the ellipse (see **Figure 15**, left plot).

2. The Sun-planet line sweeps out equal areas in the equal times. This means that planets move faster when they are nearer the Sun (perihelion) and slower when they are further away (aphelion).

3. The square of the orbital period of a planet is directly proportional to the cube of the semi-major axis *a* of its orbit. This law defines the Sun-Earth distances at any point of the orbit.

The Sun is located in one of the two foci (point -C in **Figure 15**) of the ellipse with a semi-major axis $a$ and a semi-minor axis $b$ and the eccentricity:

$$e = \sqrt{1 - \frac{b^2}{a^2}}, \tag{4}$$

It can be noted that is a link between the semi-major and semi-minor axis:

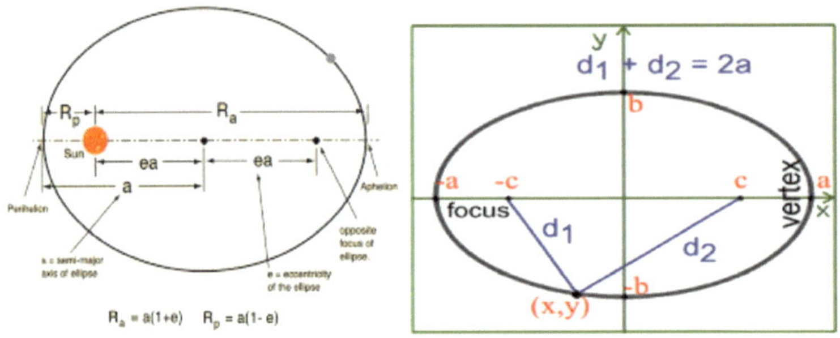


**Figure 15.**
*Elliptic orbit of the earth revolution about the sun located in the left focus of the ellipse with a semi-major axis a, a semi-minor axis b and the eccentricity e.*

$$b^2 = a^2(1 - e^2). \tag{5}$$

The planet is located in the point (x,y) on the orbit on a distance $d$ from the focus C under the angle $\theta$ to the major axis (see **Figure 15**, right plot).

This distance $d$ is defined by Kepler's third law as follows:

$$d = \frac{a(1 - e^2)}{1 - ecos\theta}, \tag{6}$$

where $0 \leq e \leq 1$.

Let us introduce $R_a$ the aphelion distance from the focus where the star is located, to the longest point of the orbit along the major axis and $R_p$ - the perihelion distance from the focus to the closest point along the major axis. Using this Eq. (6) one can calculate the aphelion $R_a$ and perihelion $R_p$ distances by setting the angle $\theta$ equal to zero for aphelion and 180° for perihelion, e.g.

$$R_a = a(1 + e), \tag{7}$$

$$R_p = a(1 - e). \tag{8}$$

The sum of the distances $d_1$ and $d_2$ from a planet location on the orbit to the both foci of the ellipse is constant and equal $2a$, e.g.

$$d_1 + d_2 = 2a = R_a + R_p. \tag{9}$$

where bf d1 and $d_2$ are distances from the two foci to the current position of a planet (see **Figure 15**, left plot).

## B. Average distances of a planet from the focus of ellipse

There are three average distances of a planet from the star can be calculated averaged in: (a) the angle $\theta$; (b) time; and (c) arc length $s$ [62]. The formula (6) for the planet distance from the ellipse focus, according to Kepler's third law, can be re-written as follows:

$$d = \frac{P_e}{1 - ecos\theta}. \tag{10}$$

where $P_e = a(1 - e^2)$.

## B.1 Planet distance averaged in angle

Average distance $\overline{d}_\theta$ in angle $\theta$ is defined by the integral:

$$\overline{d}_\theta = \frac{1}{2\pi}\int_0^{2\pi} d\cdot d\theta = \frac{1}{2\pi}\int_0^{2\pi}\frac{P_e}{1-ecos\theta}d\theta = \frac{2\pi P_e}{2\pi\sqrt{\left(1-e^2 = \frac{a(1-e^2)}{\sqrt{1-e^2)}}\right.}}, \tag{11}$$

so that

$$\overline{d}_\theta = a\sqrt{1-e^2} = b, \tag{12}$$

Hence, the average distance by $\theta$ is equal to the semi-minor axis $b$.

## B.2 Planet distance averaged in time

Average in time distance $\overline{d}_t$ is defined by the integral for the period of a planet revolution T:

$$\overline{d}_t = \frac{1}{T}\int_0^T ddt, \tag{13}$$

Since according to the second Kepler's law, the radial arm of a given planet sweeps out an area at a constant rate h:

$$h = \frac{1}{2}d^2 d\theta/dt, \tag{14}$$

or

$$dt/d\theta = d^2/2h. \tag{15}$$

We know that the area of the ellipse is $Th = \pi ab$. Hence, the average distance in time can be defined as:

$$\overline{d}_t = \frac{1}{T}\int_0^T d\cdot dt = \frac{h}{\pi ab}\int_0^{2\pi} d\cdot(dt/d\theta)d\theta = \frac{h}{\pi ab}\int_0^{2\pi} d\frac{d^2}{2h}d\theta = \frac{1}{2\pi ab}\int_0^{2\pi} d^3 d\theta, \tag{16}$$

where

$$\frac{1}{2\pi ab}\int_0^{2\pi} d^3 d\theta = \frac{p^3e^3(2+e^2)\pi}{1-e^2)^{(5/2)} = b\left(3a^2-b^2\right)\pi.} \tag{17}$$

Then the average in time distance is

$$\overline{d}_t = \frac{1}{T}\int_0^T ddt = \frac{b\left(3a^2-b^2\right)\pi}{2\pi ab} = 3a/2 - b^2/2a = \frac{3a}{2} - \frac{a^2(1-e^2)}{2a} = a\left(1+\frac{e^2}{2}\right). \tag{18}$$

or the averaged by time distance is

$$\overline{d}_t = a\left(1 + \frac{e^2}{2}\right), \tag{19}$$

which is larger than the semi-major axis a by a factor of $(1 + e^2/2)$.

## B.3 Average distance in arc length

The integral for the arc length of an ellipse cannot be evaluated in finite terms, we need to proceed indirectly utilising the defining property of an ellipse that the sum of the distances from any point of the ellipse for the two foci is constant as described by Eq. (9). Let $L$ to be the whole length of the orbit.

From the point of symmetry:

$$\int_0^L d_1 ds = \int_0^L d_2 ds. \tag{20}$$

Since by the definition of the ellipse $d1 + d2 = 2a$ (see Eq. (9), hence:

$$\int_0^L (d_1 + d_2) ds = 2aL, or \tag{21}$$

$$\int_0^L d \cdot ds = aL. \tag{22}$$

Then the distance $r_s$ averaged by the arc length is given by the expression:

$$\overline{d}_s = \frac{1}{L}\int_0^L d \cdot ds = aL/L = a. \tag{23}$$

Therefore, for the ideal revolution of a planet in ellipse about the star located in the ellipse focus, the average distance of the planet from the focus is defined by the parameters of the ellipse, along which the planet moves about the assumed location of the star if there are no other gravitational effects are considered.

## C. Average distances versus aphelion/perihelion variations.

The first step in investigation of millennial variations of solar irradiance and baseline magnetic field came from a suggestion of changing Sun-Earth distances because of a change of the Earth orbit shape. In fact, the ephemeris show [48] that since 1600 the Earth orbit's aphelion is found steadily decreasing while its perihelion is increasing (see **Figure 16**). Could this change of the Earth orbit cause the millennial changes of solar irradiance and magnetic field baseline? Let us explore this option. Sun is located in the ellipse focus, the S–E distance variations are affected by the variations of the Earth orbit parameters: aphelion and perihelion distances which are calculated by JPL. The link between the semi-minor axis $b$ and the semi semi-major axis $a$ can be written as follows:

$$2b = \sqrt{(d_1 + d_2)^2 - f^2}; \tag{24}$$

where $d_1$ and $d_2$ are the distances from the two ellipse foci to any point on the orbit, f is the distance between the foci of the ellipse, e.g. $f = R_a - R_p$. For

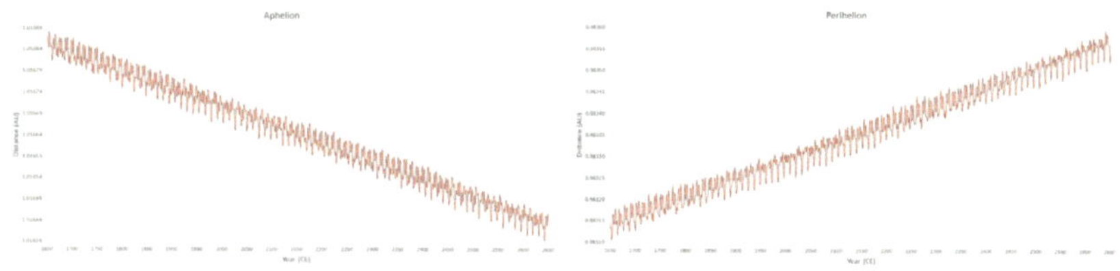

**Figure 16.**
*Variations of the aphelion (left) and perihelion (right) distances of the Earth orbit in the millennium M2.*

calculation of $d_1 + d_2$ we use the Eq. (9) that provides the relations for the semi-major $a$ and the semi-major $b$ distances via the aphelion $R_a$ and perihelion $R_p$ as follows:

$$2b = \sqrt{(d_1 + d_2)^2 - f^2} = 2\sqrt{R_a R_p} \tag{25}$$

$$a = \frac{R_a + R_p}{2}. \tag{26}$$

Hence, if we fix for some time $t_0$ the aphelion $R_{a0}$ and perihelion $R_{p0}$ distances, and assume that they are proportionally changed after some time by a magnitude $\Delta$, so that $R_a = R_{a0} - \Delta$ and $R_p = R_{p0} + \Delta$, so that $R_a + R_p = R_{a0} + R_{p0}$ and $f = R_{a0} - R_{p0} - 2\Delta$. Then from Eq. (24) the semi-minor axis can be calculated as follows:

$$\begin{aligned} 2b &= \sqrt{(d_1 + d_2)^2 - f^2} = \sqrt{\left(R_{a0} + R_{p0}\right)^2 - \left(R_{a0} - R_{p0} - 2\Delta\right)^2} = \\ &= \sqrt{4R_{a0}R_{p0} + 4\Delta\left[R_{a0} - R_{p0} - \Delta\right]}. \end{aligned} \tag{27}$$

Giving the relationships between the the ellipse axises $a$ and $b$ and aphelion and perihelion distances.

$$a = R_a + R_b = R_{a0} + R_{p0}; \tag{28}$$

$$b = \sqrt{R_{a0}R_{p0} + \Delta\left[R_{a0} - R_{p0} - \Delta\right]}. \tag{29}$$

It is evident from the equations above that variations of the aphelion and perihelion distances will affect only the Earth semi-minor axis $b$ and, thus, eccentricity $e$. Let us use them for estimation of the Earth orbit parameters from the ephemeris of aphelion and perihelion presented in **Figure 16**.

In the case of decreasing aphelion and increasing perihelion distances for a elliptic orbit with the star in its focus shown in **Figure 16**, it occurs from Eq. (29) that $b$ will increase, while the orbit eccentricity $e$ would decrease. By comparing the variations of the Earth aphelion and perihelion distances in 1500–2500 we evaluated the following changes. The variations of the aphelion and perihelion distances produce the reduction of eccentricity from 0.0170 in 1500 to 0.0163 in 2500. This would lead to a change of the average-by-time Sun-Earth distance from 1.0001462 au in 1600 to 1.0001328 au in 2500, e.g. the difference is virtually negligible and cannot be reflected in a noticeable change of solar irradiance. This indicates that over the whole millennium 1600–2600 the Earth orbit remains, in fact, a pretty stable elliptic orbit. However, the average S-E distances in this elliptic orbit do not change to such the extent to produce noticeable variations of solar irradiance or magnetic field baseline.

## Author details

Valentina Zharkova
Department of Mathematics, Physics and Electrical Engineering, Faculty of Engineering and Environment, Northumbria University, Newcastle Upon Tyne, United Kingdom

*Address all correspondence to: valja46@googlemail.com

## References

[1] E. N. Parker. Hydromagnetic Dynamo Models. Astrophysical Journal, 122:293, September 1955.

[2] Rudolf Wolf. Studies on the frequency of Sun-spots, and on their connexion with the Magnetic Declination-variation. Mon. Not. of the Royal Astron. Soc., 30:157, April 1870.

[3] W. Maunder, Mrs. Wolfer's relative sun-spot numbers. Mon. Not. of the Royal Astron. Soc., 64:226−+, January 1904.

[4] H. W. Babcock. The Topology of the Sun's Magnetic Field and the 22-YEAR Cycle. Astrophysical Journal, 133:572, March 1961.

[5] M. Stix. Differential rotation and the solar dynamo. Astronomy and Astrophysics, 47:243–254, March 1976.

[6] S. Zharkov, E. Gavryuseva, and V. Zharkova. The Observed Long- and Short-Term Phase Relation between the Toroidal and Poloidal Magnetic Fields in Cycle 23. Solar Physics, 248:339–358, April 2008.

[7] V. V. Zharkova, S. J. Shepherd, and S. I. Zharkov. Principal component analysis of background and sunspot magnetic field variations during solar cycles 21-23. Mon. Not. of the Royal Astron. Soc., 424:2943–2953, August 2012.

[8] V. Zharkova, E. Popova, S. Shepherd, and S. Zharkov. Reply to comment on the paper " on a role of quadruple component of magnetic field in defining solar activity in grand cycles" by Usoskin (2017). Journal of Atmospheric and Solar-Terrestrial Physics, 2018.

[9] S. J. Shepherd, S. I. Zharkov, and V. V. Zharkova. Prediction of Solar Activity from Solar Background Magnetic Field Variations in Cycles 21-23. Astrophysical Journal, 795:46, November 2014.

[10] V. V. Zharkova, S. J. Shepherd, E. Popova, and S. I. Zharkov. Heartbeat of the Sun from Principal Component Analysis and prediction of solar activity on a millennium timescale. Nature Scientific Reports, 5:15689, October 2015.

[11] E. Popova, V. Zharkova, and S. Zharkov. Probing latitudinal variations of the solar magnetic field in cycles 21-23 by Parker's Two-Layer Dynamo Model with meridional circulation. Annales Geophysicae, 31:2023–2038, November 2013.

[12] J. Zhao, R. S. Bogart, A. G. Kosovichev, T. L. Duvall, Jr., and T. Hartlep. Detection of Equatorward Meridional Flow and Evidence of Double-cell Meridional Circulation inside the Sun. Astrophysical Journal, Letters, 774:L29, September 2013.

[13] S. Zharkov, V. V. Zharkova, and S. S. Ipson. Statistical Properties Of Sunspots In 1996 2004: I. Detection, North South Asymmetry And Area Distribution. Solar Physics, 228:377–397, May 2005.

[14] B. Belucz and M. Dikpati. Role of Asymmetric Meridional Circulation in Producing North-South Asymmetry in a Solar Cycle Dynamo Model. Astrophysical Journal, 779:4, December 2013.

[15] J. Shetye, D. Tripathi, and M. Dikpati. Observations and Modeling of North-South Asymmetries Using a Flux Transport Dynamo. Astrophysical Journal, 799:220, February 2015.

[16] J. A. Eddy. The Maunder Minimum. Science, 192:1189–1202, June 1976.

[17] V. V. Zharkova, S. J. Shepherd, E. Popova, and S. I. Zharkov. Reinforcing a

Double Dynamo Model with Solar-Terrestrial Activity in the Past Three Millennia. In Space Weather of the Heliosphere: Processes and Forecasts, 2018.

[18] S. K. Solanki, I. G. Usoskin, B. Kromer, M. Schüssler, and J. Beer. Unusual activity of the Sun during recent decades compared to the previous 11,000 years. Nature, 431: 1084–1087, October 2004.

[19] V. V. Zharkova, S. J. Shepherd, S. I. Zharkov, and E. Popova. RETRACTED ARTICLE: Oscillations of the baseline of solar magnetic field and solar irradiance on a millennial timescale. Scientific Reports, 9:9197, June 2019.

[20] L. E. A. Vieira, S. K. Solanki, N. A. Krivova, and I. Usoskin. Evolution of the solar irradiance during the Holocene. Astronomy and Astrophysics, 531:A6, July 2011.

[21] F. Steinhilber, J. Beer, and C. Fröhlich. Total solar irradiance during the Holocene. Geophysics Research Letters, 36(19):L19704, October 2009.

[22] F. Steinhilber, J. A. Abreu, J. Beer, I. Brunner, M. Christl, H. Fischer, U. Heikkila, P. W. Kubik, M. Mann, K. G. McCracken, H. Miller, H. Miyahara, H. Oerter, and F. Wilhelms. 9,400 years of cosmic radiation and solar activity from ice cores and tree rings. Proceedings of the National Academy of Science, 109 (16):5967–5971, April 2012.

[23] Paul D. Jose. Sun's motion and sunspots. Astronomical journal, 70:193, April 1965.

[24] R. W. Fairbridge and J. H. Shirley. Prolonged minima and the 179-yr cycle of the solar inertial motion. Solar Physics, 110:191–210, March 1987.

[25] I. Charvatova. The solar motion and the variability of solar activity. Advances in Space Research, 8(7):147–150, January 1988.

[26] James H. Shirley, Kenneth R. Sperber, and Rhodes W. Fairbridge. Sun's Inertial Motion and Luminosity. Solar Physics, 127(2):379–392, June 1990.

[27] Milan Paluš, Jürgen Kurths, Udo Schwarz, Norbert Seehafer, Dagmar Novotná, and Ivanka Charvátová. The solar activity cycle is weakly synchronized with the solar inertial motion. Physics Letters A, 365(5–6): 421–428, June 2007.

[28] R. Mackey. Rhodes Fairbridge and the idea that the solar system regulates the Earth's climate. Journal of Coastal Research, (Proceedings of the Ninth International Coastal Symposium, Gold Coast, Australia), 51, May 2007.

[29] I. Charvátová. Can origin of the 2400-year cycle of solar activity be caused by solar inertial motion? Annales Geophysicae, 18(4):399–405, April 2000.

[30] A. S. Perminov and E. D. Kuznetsov. Orbital Evolution of the Sun-Jupiter-Saturn-Uranus-Neptune Four-Planet System on Long-Time Scales. Solar System Research, 52(3):241–259, May 2018.

[31] Judith Lean, Juerg Beer, and Raymond Bradley. Reconstruction of solar irradiance since 1610: Implications for climate change. Geophysics Research Letters, 22(23):3195–3198, January 1995.

[32] Charles L. Wolff and John R. Hickey. Solar Irradiance Change and Special Longitudes Due to $\gamma$-Modes. Science, 235(4796):1631–1633, March 1987.

[33] III Lee, Robert B., M. Alan Gibson, Robert S. Wilson, and Susan Thomas. Long-term total solar irradiance

variability during sunspot cycle 22. Journal of Geophysics Research, 100 (A2):1667–1676, February 1995.

[34] N. A. Krivova, S. K. Solanki, and Y. C. Unruh. Towards a long-term record of solar total and spectral irradiance. Journal of Atmospheric and Solar-Terrestrial Physics, 73(2–3):223–234, February 2011.

[35] S.-I. Akasofu. On the recovery from the little ice age. Natural Science, 2: 1211–1224, November 2010.

[36] D. J. Easterbrook. Evidence-based Climate Science. Elsevier, May 2016.

[37] Judith Lean. Evolution of the Sun's Spectral Irradiance Since the Maunder Minimum. Geophysics Research Letters, 27(16):2425–2428, August 2000.

[38] Y. M. Wang, J. L. Lean, and Jr. Sheeley, N. R. Modeling the Sun's Magnetic Field and Irradiance since 1713. Astrophysical Journal, 625(1):522–538, May 2005.

[39] P.J. *et al* Reimer. INTCAL09 AND MARINE09 RADIOCARBON AGE CALIBRATION CURVES 0–50,000 YEARS CAL BP. Radiocarbon, 51:1111–1150, 2009.

[40] J. D. Hays, John Imbrie, and N. J. Shackleton. Variations in the Earth's Orbit: Pacemaker of the Ice Ages. Science, 194(4270):1121–1132, December 1976.

[41] M. Milankovich. Canon of Insolation and the Ice Age Problem. *Zavod*za Udzbenike i Nastavna Sredstva, pages 307–310, April 1998.

[42] L. E. A. Vieira, S. K. Solanki, N. A. Krivova, and I. Usoskin. VizieR Online Data Catalog: Evolution of solar irradiance during Holocene (Vieira+, 2011). VizieR Online Data Catalog, pages J/A+A/531/A6, May 2011.

[43] I. G. Usoskin, Y. Gallet, F. Lopes, G. A. Kovaltsov, and G. Hulot. Solar activity during the Holocene: the Hallstatt cycle and its consequence for grand minima and maxima. Astronomy and Astrophysics, 587:A150, March 2016.

[44] N. Scafetta. Discussion on the spectral coherence between planetary, solar and climate oscillations: a reply to some critiques. Astrophysics and Space Science, 354:275–299, December 2014.

[45] P. Stauning. Solar activity-climate relations: A different approach. Journal of Atmospheric and Solar-Terrestrial Physics, 73(13):1999–2012, August 2011.

[46] Rodolfo G. Cionco and Dmitry A. Pavlov. Solar barycentric dynamics from a new solar-planetary ephemeris. Astronomy and Astrophysics, 615:A153, July 2018.

[47] P. Bretagnon and G. Francou. Planetary Theories in rectangular and spherical variables: VSOP87 solution. Astronomy and Astrophysics, 202:309, August 1988.

[48] W. M. Folkner, J. G. Williams, D. H. Boggs, R. S. Park, and P. Kuchynka. The Planetary and Lunar Ephemerides DE430 and DE431. Interplanetary Network Progress Report, 42-196:1–81, February 2014.

[49] Duncan Steel. Perihelion precession, polar ice and global warming. Journal of Cosmology, 22:10106–10129, March 2013.

[50] V. S. Airapetian, R. Barnes, O. Cohen, G. A. Collinson, W. C. Danchi, C. F. Dong, A. D. Del Genio, K. France, K. Garcia-Sage, A. Glocer, N. Gopalswamy, J. L. Grenfell, G. Gronoff, M. Güdel, K. Herbst, W. G. Henning, C. H. Jackman, M. Jin, C. P. Johnstone, L. Kaltenegger, C. D. Kay, K. Kobayashi, W. Kuang, G. Li, B. J. Lynch, T. Lüftinger, J. G. Luhmann, H. Maehara,

M. G. Mlynczak, Y. Notsu, R. A. Osten, R. M. Ramirez, S. Rugheimer, M. Scheucher, J. E. Schlieder, K. Shibata, C. Sousa-Silva, V. Stamenković, R. J. Strangeway, A. V. Usmanov, P. Vergados, O. P. Verkhoglyadova, A. A. Vidotto, M. Voytek, M. J. Way, G. P. Zank, and Y. Yamashiki. Impact of space weather on climate and habitability of terrestrial-type exoplanets. International Journal of Astrobiology, 19(2):136–194, April 2020.

[51] Kevin France. Exploring extreme exoplanets. Nature Astronomy, 4:1112–1112, November 2020.

[52] V. V. Zharkova, S. J. Shepherd, and E. Popova. Oscillations of the baseline of solar magnetic field and solar irradiance on a millennial timescale. arXiv e-prints, page arXiv:2002.06550, February 2020.

[53] D. Halliday, R. Resnick, and J. Walker. Fundamentals of Physics: Extended. Wiley, Germany, 2010.

[54] Richard C. Willson and Hugh S. Hudson. The Sun's luminosity over a complete solar cycle. Nature, 351(6321): 42–44, May 1991.

[55] M. Fligge and S. K. Solanki. The solar spectral irradiance since 1700. Geophysics Research Letters, 27(14): 2157–2160, August 2000.

[56] M. Lockwood and R. Stamper. Long-term drift of the coronal source magnetic flux and the total solar irradiance. Geophysics Research Letters, 26(16):2461–2464, January 1999.

[57] Drew T. Shindell, Gavin A. Schmidt, Michael E. Mann, David Rind, and Anne Waple. Solar Forcing of Regional Climate Change During the Maunder Minimum. Science, 294(5549):2149–2152, December 2001.

[58] Gifford H. Miller, Áslaug Geirsdóttir, Yafang Zhong, Darren J. Larsen, Bette L. Otto-Bliesner, Marika M. Holland, David A. Bailey, Kurt A. Refsnider, Scott J. Lehman, John R. Southon, Chance Anderson, Helgi Björnsson, and Thorvaldur Thordarson. Abrupt onset of the Little Ice Age triggered by volcanism and sustained by sea-ice/ocean feedbacks. Geophysics Research Letters, 39(2): L02708, January 2012.

[59] Hermann Harde. Scrutinizing the carbon cycle and $CO_2$ residence time in the atmosphere. Global and Planetary Change, 152:19–26, May 2017.

[60] Valentina Zharkova. Modern grand solar minimum will lead to terrestrial cooling. Temperature, 7(3):217–222, 2020.

[61] Johann Kepler. Astronomia nova. 1609.

[62] S. K. Stein and T. Elsner. Mean Distance" in Kepler's Third Law. Mathematics Magazine, 50(3):160–162, March 1977.

Section 3

# Venus

Chapter 3

# Solar Cycle Variations in the Position of Vortex Structures in the Venus Wake

*H. Pérez-de-Tejada and R. Lundin*

## Abstract

Measurements conducted with the Venus Express (VEX) spacecraft at its entry and exit through vortex structures in the Venus wake reveal that their position varies with the solar cycle. Both crossings are consistently measured closer to Venus during minimum solar cycle conditions and are gradually encountered at larger distances downstream from the planet along the solar cycle. At the same time their width along the VEX trajectory on the plane transverse to the solar wind direction is larger during minimum solar cycle conditions and show a gradual decrease along the solar cycle. As a result the vortex structures are envisioned as features that gradually become thinner as they extend along the Venus wake and agree with the geometry of a vortex flow in fluid dynamics whose thickness decreases with the downstream distance from an obstacle. Similar conditions should also be applicable to Mars and other bodies within the solar system and also possibly to exo-planets in external stellar systems.

**Keywords:** Venus Wake, Vortex Structures, Pressure Balance Conditions

## 1. Introduction

Measurements conducted with various spacecraft across the Venus wake: the Mariner 5, the Venera, the Pioneer Venus Orbiter (PVO), and the Venus Express (VEX) spacecraft have provided evidence of plasma features that resemble vortex structures in fluid dynamic problems. It has been noted that the anti-solar directed flow of the plasma particles in the Venus wake is rotated to even become oriented back to Venus. The early measurement of that effect was reported by Pérez-de-Tejada et al. [1–3] from the PVO plasma data and that gave place to conditions that could be assessed as resulting from a vortex structure. Further information on the observation of vortex structures in the Venus wake was reported by Lundin et al. [4, 5] from measurements conducted with the ASPERA instrument onboard VEX. In this case it was possible to identify the scale size and position of vortex structures with information on the distribution of the velocity vectors of the solar wind and planetary ions that stream within and around those features.

A general view of the shape of a vortex structure is reproduced in the left panel of **Figure 1** showing in the plane transverse to the solar wind direction a vortex structure with a scale size comparable to that of the planet with a circulation sense that is counterclockwise when seen from the wake back to Venus. A circulation

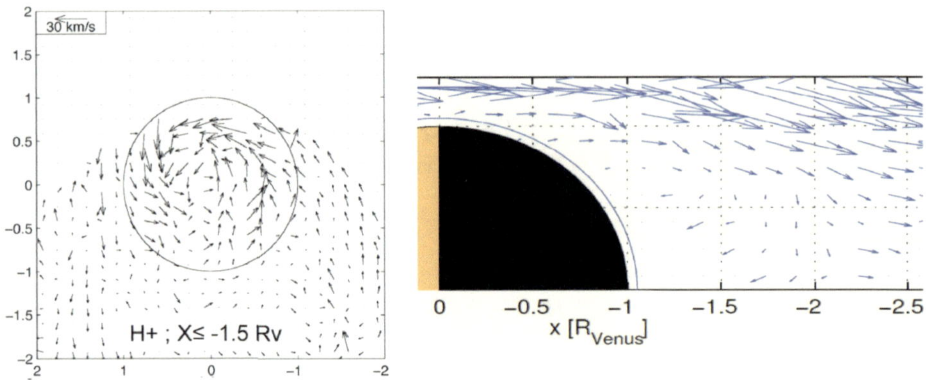


**Figure 1.**
*(left panel) velocity vectors of H+ ≈ 1–300 eV ions measured with the VEX spacecraft in the Venus wake projected on the YZ plane transverse to the solar wind direction. Data are averaged in 1000 x 1000 km columns at X < −1.5 $R_V$ (adapted from **Figure 4** of [5]). (right panel) average direction of solar wind ion velocity vectors across the Venus near wake collected from many VEX orbits and projected in cylindrical coordinates [15].*

pattern of the solar wind direction in cylindrical coordinates along the Venus wake is added in the right side panel of **Figure 1** with indications of a sunward directed flow return in the central wake. Comparable variations in the plasma velocity vector have also been recently reported from the VEX measurements [6]. Different from that motion there has also been information derived from the PVO and VEX measurements on an east–west displacement of planetary ion fluxes and that leads to the overall deflection of the trans-terminator ionospheric flow toward the dawn-hemisphere as it moves into the night-side ([7], see Fig. 15; [4], see Fig. 7b). A deflection in that direction can be accounted for in terms of the fluid dynamic Magnus force produced by the joint contribution of the unidirectional solar wind velocity and the rotation of the Venus atmosphere/ionosphere [8, 9].

## 2. Vortex structures

Ample information on the speed and density values of the solar wind and planetary ions is their energy spectra obtained with the ASPERA instrument onboard the VEX spacecraft. A suitable example is provided by the data of the Sept. 26–2009 orbit reproduced in **Figure 2** and that probed by the near vicinity of the midnight plane (small Y-values shown at the bottom of the figure). The energy spectra of the O+ ion component (second panel) exhibit variations that indicate the presence of appreciable planetary O+ ion fluxes between 02:05 UT and 02:30 UT and that lead to enhanced values of their density and speed (third and fifth panels). At the same time the magnetic field intensity exhibits decreased values within that time interval with an oscillating response of its components (seventh panel). The later indicates the possible presence of a vortex structure within the region where enhanced values of the density and speed of the O+ ion fluxes are clearly distinguished. An important property of the observed values is that the kinetic energy density of such fluxes ($mnv^2 \sim 10\ 10^{-10}$ ergs $cm^{-3}$) is comparable to the magnetic energy density ($B^2/8\pi \sim 4\ 10^{-10}$ ergs $cm^{-3}$) measured in the vicinity of that region where B ≈ 10 nT. As a result a near pressure balance condition between values measured outside is suggested within a region with evidence of a vortex structure. It is to be noted that these variations occurred far downstream from the planet (by X < −1.70 $R_V$ with this latter parameter being the Venus radius) and thus are unrelated to the crossing of the VEX

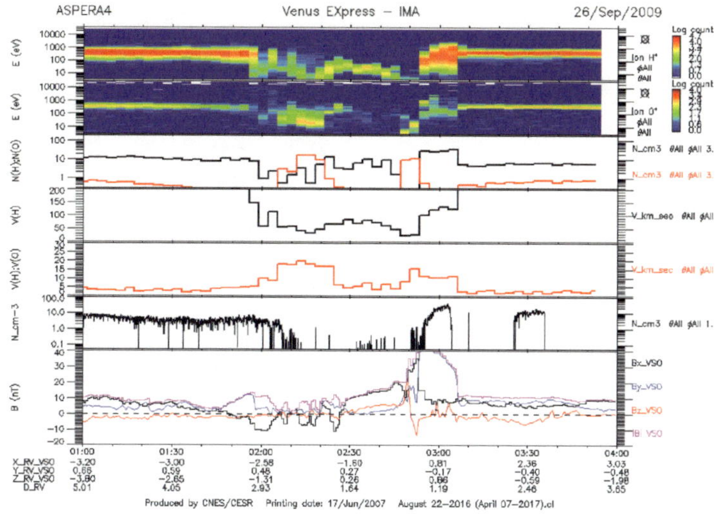


**Figure 2.**
*Energy spectra of the H+ and O+ ions (upper panels) measured during the sept 26–2009 VEX orbit in the Venus wake by the midnight plane (small Y-values below the figure). In measurements between 02:05 UT and 02:30 UT there are oscillations in the magnetic field components (bottom panel) indicating a vortex structure. In that time interval there are also decreased values of the magnetic field intensity with enhanced O+ density and speed values (third and fifth panels).*

spacecraft near the Venus ionosphere where $X \approx 0$ (X, Y, and Z represent Cartesian coordinates indicating, respectively, the Venus-sun direction, the direction opposite to the motion of Venus around the sun, and the axis transverse to the plane formed by both vectors).

Similar conditions have also been identified in other VEX orbits that were traced by the vicinity of the midnight plane and that exhibit as well a near pressure balance condition between the region where enhanced density and speed values of the planetary ions are measured with the magnetic energy density of the magnetic fluxes encountered in their vicinity [10]. At the same time there is no evidence of a sudden and strong reversal in the direction of the $B_x$ magnetic field component as it would be expected across a plasma sheet embedded by the middle of the wake. On the contrary the enhanced values of the density of the O+ ion population measured by the vortex structure indicates that there is not a local plasma expansion in that region.

## 3. Distribution of vortex structures in the Venus wake

Different from those properties it is necessary to examine changes of the vortex structures in their position along the nearly 8 years of observations conducted with the VEX spacecraft (between 2006 and 2013). A quantitative analysis of their location was made when VEX entered and exited those structures that was derived from the energy spectra of the planetary ions in the 20 VEX orbits listed in **Table 1**. A comparative view of the distribution of the vortex structures on the XZ plane of the solar wind velocity direction is presented in **Figure 3** to show the position of the VEX entry and exit crossings in orbits that probed near the midnight plane. Two sets with 4 orbits corresponding to measurements made in 2006 and in 2009 indicate a different displacement of the vortex structures in the Z-direction. There is a general preference of those features to occur closer to Venus in the 2009 measurements since their passage across the $Z = 0$ axis is by $X = -1.7\ R_V$ in that set while it reaches $X = -2.2\ R_V$ in the 2006 measurements.

| Date | UT | X | Y | Z | n | v | $(\rho v^2)$ | ΔT |
|---|---|---|---|---|---|---|---|---|
| Aug 22-2006 | 01:45 | -2.87 | -0.15 | -1.12 | 10 | 20 | 10 | 9 |
| Aug 22-2006 | 01:54 | -2.55 | -0.15 | -0.40 | 10 | 20 | 10 | 9 |
| Aug 23-2006 | 01:59 | -2.64 | -0.07 | -1.30 | 10 | 15 | **6** | 6 |
| Aug 23-2006 | 02:05 | -2.40 | -0.07 | -0.44 | 10 | 15 | **6** | **6** |
| Aug 24-2006 | 02.10 | -2.48 | 0.01 | -0.83 | 10 | 30 | 23 | 10 |
| Aug 24-2006 | 02:20 | -2.04 | 0.01 | 0.20 | 10 | 30 | 23 | 10 |
| Aug 28-2006 | 02:22 | -2.40 | 0.28 | -0.38 | 20 | 20 | 20 | 6 |
| Aug 28-2006 | 02:28 | -2.18 | 0.07 | 0.91 | 20 | 20 | 20 | 6 |
| Nov 13-2007 | 00:56 | -2.60 | -0.22 | -0.70 | 10 | 20 | 10 | 7 |
| Nov 13-2007 | 01:03 | -2.38 | -0.21 | -0.38 | 10 | 20 | 10 | 7 |
| Nov 15-2007 | 00:57 | -2.60 | -0.08 | -0.65 | 10 | 20 | 10 | 6 |
| Nov 15-2007 | 01:03 | -2.40 | -0.08 | -0.38 | 10 | 20 | 10 | 6 |
| June 27-2008 | 03:26 | -2.60 | 0.03 | -0.14 | 5 | 15 | **3** | 10 |
| June 27-2008 | 03:36 | -2.30 | 0.04 | -0.20 | 5 | 15 | **3** | 10 |
| June 28-2008 | 03:33 | -2.52 | 0.04 | -0.14 | 5 | 15 | **3** | 12 |
| June 28-2008 | 03:45 | -1.96 | 0.01 | -0.20 | 5 | 15 | **3** | 12 |
| Sept 19-2009 | 01:54 | -2.42 | -0.04 | -1.04 | 10 | 15 | 6 | 9 |
| Sept 19-2009 | 02:03 | -2.11 | -0.05 | -0.55 | 10 | 15 | 6 | 9 |
| Sept 21-2009 | 02:02 | -2.30 | 0.08 | -0.65 | 10 | 15 | **6** | 10 |
| Sept 21-2009 | 02:12 | -1.95 | 0.06 | -0.12 | 10 | 15 | **6** | 10 |
| Sept 25-2009 | 02:14 | -2.15 | 0.33 | -0.45 | 10 | 20 | 10 | 13 |
| Sept 25-2009 | 02:27 | -1.60 | 0.23 | 0.21 | 10 | 20 | 10 | 13 |
| Sept 26-2009 | 02:12 | -2.30 | 0.42 | -0.70 | 10 | 20 | 10 | 10 |
| Sept 26-2009 | 02:22 | -1.95 | 0.34 | -0.20 | 10 | 20 | 10 | 10 |
| Aug 22-2010 | 08:24 | -1.80 | 0.01 | 0.25 | 6 | 25 | 10 | 7 |
| Aug 22-2010 | 08:31 | -2.08 | 0.01 | 0.04 | 1 | 18 | 8 | 7 |
| Aug 23-2010 | 08:16 | -1.25 | 0.05 | 0.68 | 10 | 15 | 6 | 13 |
| Aug 23-2010 | 08:29 | -2.02 | 0.05 | 0.02 | 8 | 7 | 10 | 13 |
| July 23-2011 | 02:53 | -2:53 | -0.14 | -0.90 | 2 | 18 | 16 | 13 |
| July 23-2011 | 03:06 | -1.93 | -0.13 | -0.40 | 2 | 12 | 7 | 13 |
| July 29-2011 | 03:07 | -2.25 | -2.25 | -1.00 | 10 | 20 | 10 | 16 |
| July 29-2011 | 03:23 | -1.70 | 0.17 | -0.10 | 7 | 15 | 4 | 16 |
| Mar 05-2012 | 06:23 | -2.07 | -0.04 | -0.60 | 30 | 25 | 48 | 12 |
| Mar 05-2012 | 06:35 | -1.50 | -0.05 | 0.15 | 30 | 20 | 31 | 12 |
| Oct 16-2012 | 01:26 | -2.00 | -0.02 | -0.85 | 2 | 28 | 4 | 13 |
| Oct 16-2012 | 01:39 | -1.70 | -0.03 | 0.10 | 6 | 13 | 26 | 13 |
| May 29-2013 | 04:28 | -1.70 | -0.01 | -0.14 | 8 | 20 | 8 | 10 |

| Date | UT | X | Y | Z | n | v | ($\rho v^2$) | ΔT |
|---|---|---|---|---|---|---|---|---|
| May 29-2013 | 04:38 | -1.28 | -0.02 | 0.40 | 3 | 20 | 3 | 10 |
| May 30-2013 | 04:20 | -2.00 | 0.07 | -0.73 | 2 | 15 | 1 | 10 |
| May 30-2013 | 04:30 | -1.68 | -0.04 | -0.12 | 1 | 10 | <1 | 10 |

**Table 1.**
*VEX coordinates (in $R_V$) at the time of its crossing (in UT) during an entry and exit (third to fifth columns) through a plasma structure within the Venus wake. Values for the density (cm$^{-3}$) and speed v (km/s) of planetary O+ ions together with their kinetic energy density $\rho v^2$ ($10^{-10}$ ergs/cm$^{-3}$) are given in the sixth to eight columns, and the last column has the time width of the vortex structure measured between the inbound and the outbound crossings (the segment between x = −2.30 $R_V$ and x = −1.95 $R_V$ is the same for both the sept 21–2009 and the sept 26–2009 orbits).*

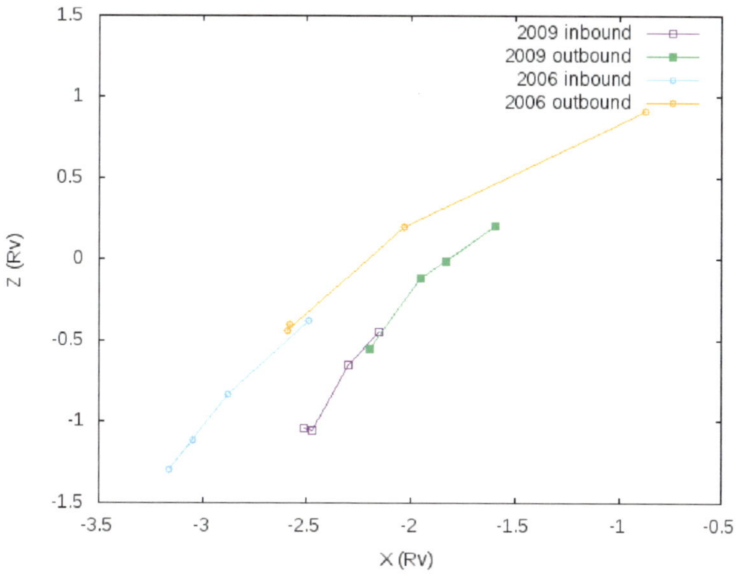


**Figure 3.**
*Position of the VEX spacecraft projected on the XZ plane during its entry (inbound) and exit (outbound) through a corkscrew plasma structure in orbits traced by the midnight plane. The two traces correspond to 4 orbits in 2006 and 2009 [10].*

This difference implies that the vortex structures are located closer to Venus during solar cycle minimum conditions by 2009 and that their position along the wake varies during that cycle.

A more extended description of the position and geometry of the vortex structures implied by the data of the orbits listed in **Table 1** is depicted in **Figure 4** to show changes in their location and extent during 8 years of VEX operation. In particular, they describe variations in their width as follows: For each orbit there is a segment bounded by the entry and exit of the spacecraft with a number that marks the two last digits of the year when measurements were made (they include the 4 orbits for 2006 and 4 for 2009 that were discussed in **Figure 3**). Most notable is that the segments identify 2 different regions; one corresponding to orbits before the minimum solar cycle conditions (between 2006 and 2009) and the other to orbits that occurred during and after those conditions (between 2009 and 2013). Two big circles select schematically different set of orbits that are located either far away from Venus between 2006 and 2009 (left circle) and

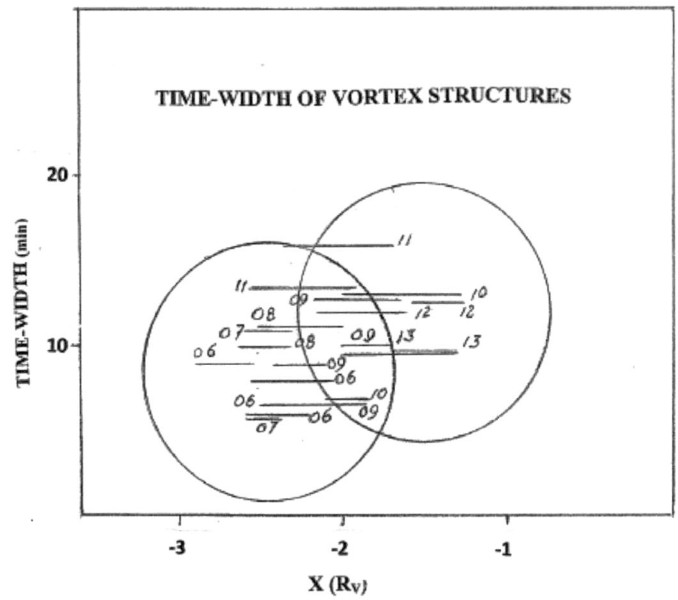


**Figure 4.**
*Time-width (in minutes) measured between the VEX inbound and outbound crossings of a vortex structure as a function of the X-distance ($R_V$) downstream from Venus in 20 orbits. The numbers at the side of each segment represent the two last digits of the year when measurements were made in different orbits between 2006 and 2013 (four orbits were examined during 2006 and 2009). The two circles confine orbits between 2006 and 2009 (left circle) prior to a solar cycle minimum and those between 2009 and 2013 (right circle) during that period.*

those that are placed closer to Venus between 2009 and 2013 (right circle) during solar cycle minimum conditions. The implication here is that as in **Figure 3** vortex structures occur closer to Venus during minimum solar cycle conditions.

Equally notable is that the time width ΔT (segment length) in the 2006–2009 orbit range is clearly smaller (placed at lower values along the vertical coordinate within the left circle) than that in the 2009–2013 orbit range (larger values in the right circle). As a result the thickness of the vortex structures located far away from Venus (left circle) becomes smaller with increasing distance along the wake thus implying that like in a corkscrew flow they thin out with distance downstream from Venus.

A schematic view of a corkscrew vortex flow structure in fluid dynamics is illustrated in **Figure 5** to represent the equivalent geometry of a similar structure in the Venus wake and that is formed by the distribution of planetary $O+$ ions eroded by the solar wind from the Venus ionosphere. The shape of the vortex structure in **Figure 5** shows how it becomes thinner with increasing distance from an object that is immersed in a streaming fluid. Such representation is consistent with the wider width of the vortex structures measured closer to Venus in the 2009–2013 orbit set (right circle in **Figure 4**) during solar minimum conditions instead of their thinner width measured in the 2006–2009 orbit set measured before the minimum solar cycle conditions (left circle), and that are located farther away along the Venus wake. As a result the thickness of the vortex structure gradually decreases with distance downstream from Venus and that eventually fade away and diffuse with the solar wind plasma. Further studies of more extended data are required to examine the evolution of the vortex structures far downstream along the Venus tail. It should also be noticed that in addition to changes associated to a solar cycle

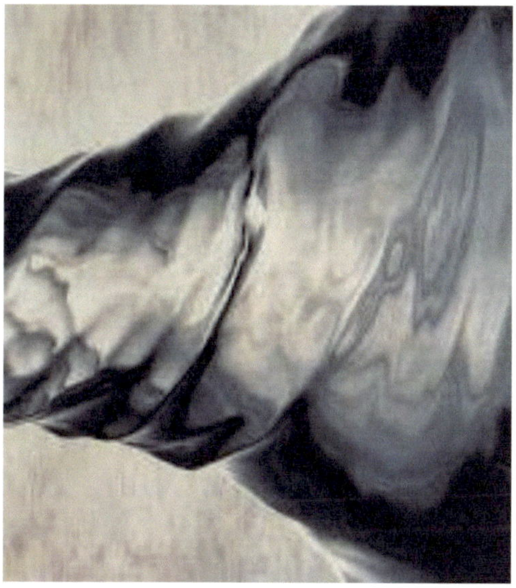

**Figure 5.**
*View of a corkscrew vortex flow in fluid dynamics. Its geometry is equivalent to that of a vortex flow in the Venus wake with its width and position varying during the solar cycle. Near minimum solar cycle conditions the vortex is located closer to Venus (right side) and there are indications that its width becomes smaller with increasing distance downstream from the planet. Such is the case for orbits within the left circle of* ***Figure 4*** *and that were conducted before the solar cycle minimum at 2009–2010 thus implying that it becomes thinner when it is detected further downstream along the wake.*

there is evidence that the solar energy output and thus that of the solar wind during the last cycle have decreased with respect to values measured in the previous cycle and thus there is a tendency for the input solar wind pressure to now reach smaller values under solar minimum conditions.

## 4. Conditions applicable to other planets

The main output of these concepts is that the position of the vortex structures along the Venus wake and their width measured along the VEX trajectory vary along the solar cycle thus implying a continuous displacement of the region where they apply. Similar conditions should also be applicable along the plasma wake of other planets and satellites within the solar system and may as well reach those suitable to exo-planets impacted by stellar winds.

Throughout the solar system the most likely candidates are Venus and Mars which do not have a strong intrinsic magnetic field that deflects the solar wind before it reaches their upper atmosphere/ionosphere. While plasma vortices have been inferred from measurements made along the flanks of the earth's magnetosphere [11–13] changes in their position are not expected to produce local variations as notable as those measured in the Venus wake. On the other hand observations of vortex oscillations have been reported from measurements conducted with the MAVEN spacecraft by the solar wind Mars ionosphere boundary [14] and it is possible that they produce vortex structures similar to those detected in the Venus wake. Comparable conditions may also occur around the Titan ionosphere when it is subject to the solar wind flow and/or to plasma motion within Saturn's magnetosphere when it moves across it. The dynamic pressure of the plasma pressure that reaches Titan is very different in both cases and thus should produce vortex structures that are situated at different locations along its plasma wake. External to the solar system it is also possible that exo-planets moving within stellar systems will be subject to stellar winds that do not have the same intensity and thus arrive from different directions. In such cases vortex structures may be present at varying distances along their wake and be located at varying distances downstream from exo-planets. Stellar winds with different flow intensity, plasma temperature, and stellar irradiance will in addition influence the position and configuration of vortex structures in an exo-planet wake. A notable effect is the subsonic to supersonic speed change that depends on the coronal temperature of a star as implied from the solution of the fluid dynamic Parker equations of solar wind acceleration [15]. Equally important is the requirement that fluid dynamic vortex structures will be influenced by a planetary magnetic field as it is the case along the earth's wake.

## 5. Conclusions

The fluid dynamic response of the solar wind that interacts with planetary ionospheres is ultimately produced by wave-particle interactions inferred from the observation of frequent oscillations in magnetic field profiles measured in the Venus wake [16–18]. As a whole such interactions provide a mechanism to transfer momentum between both ion populations and thus erode the upper layers of the Venus ionosphere [2]. That process implies in turn local plasma heating [19, see Fig. 35; 20, 21] and the removal of ionospheric plasma in the form of channels or ducts that extend downstream along the Venus wake [22, 23]. These various features are in agreement with phenomena observed in fluid dynamic problems and provide an account of measurements conducted in the Venus wake with various spacecraft. The identification of vortex structures in the Venus wake whose position and width varies during the solar cycle provides a remarkable result that justifies the importance of the fluid dynamic approach.

## Acknowledgements

We wish to thank Gilberto A. Casillas for technical work provided. Financial support was available from the INAM-IN108814-3 Project. Appreciation is given to Ann Pérez-de-Tejada for comments received.

## Author details

H. Pérez-de-Tejada[1*] and R. Lundin[2]

1 Institute of Geophysics, UNAM, Mexico

2 Swedish Space Research Institute, Kiruna, Sweden

*Address all correspondence to: hectorperezdetejada@gmail.com

## References

[1] Pérez-de-Tejada, H. et al., Plasma vortex in the Venus wake, EOS, 63(18), 368, 1982.

[2] Pérez-de-Tejada, H., R. Lundin, D. Intriligator, Plasma vortices in planetary wakes, Chapter 13 in: Open Questions in Cosmology, G. Olmo, Ed., INTECH Pub., (ISBN 978-953-51-0880-1), 2012.

[3] Pérez-de-Tejada, H., et al., Vortex structure in the plasma flow channels of the Venus wake, Chapter 1 in: Vortex Structures in Fluid Dynamic Problems, H. Pérez-de-Tejada, Ed, INTECH Pub., (ISBN 978-953-51-2943-1), 2017.

[4] Lundin, R. et al., Ion flow and momentum transfer in the Venus plasma environment, ICARUS, 215, 7, 2011.

[5] Lundin, R. et al., A large scale vorstex in the Venus plasma tail and its fluid dynamic interpretation, Geophysical Research Letters, 40(7), 273, 2013.

[6] Kollmann, P. et al., Properties of planet-ward ion flows in the Venus magnetotail, ICARUS, 2016-08, doi:10.1016/icarus;2016.02.053.

[7] Miller, K., and R. Whitten, Ion dynamics in the Venus ionosphere, Space Science Rev. 55, 165. 1991.

[8] Pérez-de-Tejada, H., Magnus force in the Venus ionosphere, Journal of Geophysical Research, 111(A 11), DOI: 10.1029/ 2005/A01/554, 2006.

[9] Pérez de Tejada, H., Viscous Magnus force for the rotating Venus ionosphere, Astrophys. Journal. Lett., 676, L65, 2008.

[10] Pérez-de-Tejada, H., and R. Lundin, Vortex-related plasma boundary in the Venus wake, ICARUS (submitted, 2021).

[11] Hones Jr., et al., Vortices in magnetospheric plama flow, Geophys. Res. Lett, 5, 1059 EOF-1062, 1978.

[12] Hones Jr., et al., Further determination of the characteristics of magnetospheric plasma vortices with ISSE 1 and 2, Journal Geophys. Research, 86, 814, EOF-820, 1981.

[13] Hones Jr., et al., New observations of plasma vortices and insights into their interpretation. Geophys. Res. Lett. 10, 674, EOF-677, 1983.

[14] Ruhunusiri, S., et al., MAVEN observations of partially developed Kelvin-Helmholtz vortices at Mars, Geophysical Research Letters,10.1002/2016GL068926, 2016.

[15] Dessler, A., Solar wind and interplanetary magnetic field, Rev. of Geophysics, Vol. 5, No. 1, 1967.

[16] Bridge, H., A. et al., Plasma and magnetic fields observed near Venus, Science,158, 1669-1673, 1967 (see Fig. 2).

[17] Reyes-Ruiz, M., E. Diaz-Mendez; H. Pérez de Tejada, Numerical simulation of the viscous interaction of the solar wind with the magnetic polar regions of Venus, Astron. and Astrophys., 489, 1319, 2008.

[18] Vörös, Z. et al., Intermittent turbulence, noisy fluctuations and wavy structures in the Venusian magnetosheath and wake, Journal of Geophysical Research,2008, 113, ED0B21, doi.1029/2008JE003159, 2008.

[19] Phillips, J., and D. McComas, The magnetosheath and magnetotail of Venus, Space Sciences Rev., 55, 1, 1991 (Fig. 35).

[20] Romanov, S. A., et al., Interaction of the solar wind with Venus, Cosmic Res.,16, 603, 1979 (Fig 5).

[21] Verigin, M. et al., Plasma near Venus from the Venera 9 and 10 wide angle

analyzer data, J. Geophys. Res., 83, 3721, 1978.

[22] Pérez-de-Tejada, H., et al., Solar wind driven plasma fluxes from the Venus ionosphere, Journal of Geophysical Research, 118, 1-10, doi:10.1002/2013JA019029, 2013.

[23] Pérez-de-Tejada, H., et al., Measurement of plasma channels in the Venus wake, ICARUS, 321, 1026-1037, 2019.

Section 4

# Mars

Chapter 4

# Sedimentation and Proposed Algorithms to Detect the Possible Existence of Vegetation and Humidity in the Landing Area of the Mars Exploration Rover-B (Opportunity)

*Emilio Ramírez-Juidías, Katherine Villavicencio-Valero and Arthur Borja*

## Abstract

Opportunity was launched in 2004 and has been providing interesting data from Mars till 2018. Meridiani Planum was the landing site for the robot. This crater has numerous rock outcrops, which are considered a valuable geological resource that contains keys to the Martian past. In this work, several algorithms have been developed for detecting the possible presence of humidity and vegetation on Mars through the images sent by the Mars Exploration Rover - B Opportunity and by the Viking Orbiter between 1976 and 1980. For this, it was carried out a sedimentary simulation of the study area, as well as an analysis of all the images from the spectral signatures extracted. The results show the existence of three types of water on the surface, as well as concentrations of Neoxanthin, also on landing area surface, that suggest the possible existence of microalgae.

**Keywords:** humidity, vegetation index, chlorophyll, Meridiani Planum

## 1. Introduction

In its remote past, Mars exhibited volcanic activity, endured numerous impacts from planetesimals on its surface, but it also possessed an Earth-like atmosphere and surface water. During Mars' formation, some collisions between the planetesimals and the protoplanet ejected material into space. Of the tens of thousands of meteorites found on Earth, about two hundred came from Mars and provide some scientific evidence as to the formation of Mars [1].

Recent discoveries of thousands of exoplanets indicate that planetary systems are the norm rather than the exception in the observable universe. The types of planets that form in a system are dependent on the age of the host star, which is why the older the star, the more likely it is for the planets to contain heavier elements.

It must be emphasized that Mars was formed through an accretion process from volatile-poor planetesimals in the inner solar nebula. The relatively small size of Mars indicates possible Jovian interference during the accretion process. The accretion process and the decay of radioisotopes released gravitational potential energy that melted the Martian protoplanet. Both events led to differentiation into a core–mantle–crust planet structure [2, 3].

It is known from Earth analog studies that most of the tungsten, iron, and other metals sink to the core of a planet in its molten state early in its formation. From radioisotope ratio studies of the Martian meteorites, evidence supports the hypothesis that Mars formed in 2 to 4 million years. However, recent research suggests that tungsten and platinum on the planetesimals themselves could have altered the Martian crust ratios of tungsten, iron, and other metals. The provenance of the two hundred meteorites found on Earth does not provide conclusive evidence of the Martian surface's composition. The latest literature suggests that it is most likely that the small sample of meteorites comes from a few impacts. Those few impacts render the sample size too small to draw accurate conclusions of Martian formation. An alternate hypothesis states that large projectiles with formed cores and mantles bombarded the red planet [4]. According to new modeling, the planetesimals' composition varied enough in ferrophilic element content to provide the tungsten variance, resulting in a heterogeneous Martian mantle. This alternate hypothesis offers an estimate of a Martian formation period of up to 20 Mya [5].

On the other hand, and in relation to the Mars atmosphere, the Mariner mission revealed a fine Martian atmosphere that varies greatly. Because it is thin, the atmosphere of this planet expands rapidly in the warmer months and contracts rapidly in the colder months [6]. In general, it can be said that the atmosphere of the red planet is similar in composition to that of Venus but much thinner. Carbon dioxide comprises about 95% of the atmosphere, with the rest being nitrogen (3%) and argon (1%). The carbon dioxide freezes over the poles, and frozen water lays underneath the frozen carbon dioxide layer.

Regarding the geology of Mars, it is interesting to comment that this planet has two vastly different hemispheres (**Figure 1**), that is, the older heavily cratered highlands of the Southern Hemisphere and the Northern Hemisphere's younger lowland plains. Mars presents extreme topography when compared to Earth and even the Moon. A marked 30-kilometer elevation difference exists between the summit of Olympus Mons and the floor of the Hellas Basin. The hemispheres' difference can be explained by possible volcanic eruptions or seas that smoothed out the Northern Hemisphere. In addition to the highland terrains and the lowland

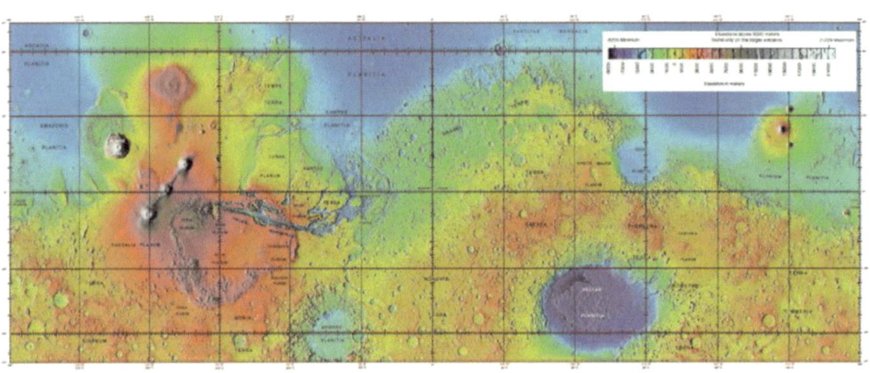

**Figure 1.**
*MOLA global image of Mars surface [7].*

plains, Mars features other landscapes that include the polar terrains, the Tharsis and Elysium volcanic terrains, and chaotic terrains.

According to [4, 5], Mars also features numerous buried impact basins dating from its early history. Other geologic features visible on the surface include dendritic runoff channels. These features display river-like patterns that evidence precipitation or snowmelt runoff flow. In the equatorial regions, outflow channels evidence cataclysmic events of the release of underground ice melt.

For the present work, and although studying Martian rocks and mineral geochemistry provides essential clues as to the conditions when they formed, it must be taken into account that it is well understood from the study of Earth analogs that chemical systems tend toward equilibrium reactions that yield the lowest energy and, therefore, the most chemically stable compounds within specific environments. As a consequence of this, it is interesting to mention that chemical weathering and erosion of basaltic and other igneous rocks produce distinct sediments and minerals that provide a detailed history in terms of temperature, pressure, pH, atmospheric composition, etc., that led to the formation of specific minerals.

Although numerous studies of various nature have been carried out on Mars to date, none of them have studied in depth the phenomena of water erosion that occur on the surface of Mars. In the same way, no scientific work has used state-of-the-art remote sensing techniques, based on machine learning, to be able to obtain algorithms capable of predicting both surface moisture content and indications of the presence of some type of photosynthetic organism. It is therefore that this chapter is innovative, in addition to pretending to be a reference in the bibliography about similar studies on rocky planets and moons in our solar system.

## 2. Study area

As is known, Opportunity (Mars Exploration Rover-B) landed in the Eagle crater, located in the Meridiani Planum (0,2° N; 357,5° E in planetocentric coordinates), on January 25, 2004.

The reason why Opportunity was sent to Meridiani Planum was because the Thermal Emission Spectrometer (TES), of the Mars Global Surveyor mission, found, from its orbit, crystalline gray hematite on the surface of Mars in an amount around 20%. Hematite is an iron oxide that, usually its gray crystalline variety, is formed in association with liquid water on Earth.

According to [8], at Opportunity's landing site, gray hematite within a kind of spherules was found in outcrops of soft and stratified sandstone rocks. It should be emphasized that for these structures to form, the acidic aqueous alteration of basalt rocks rich in goethite (a mineral that contains iron) was necessary. Subsequently, the alteration of the goethite gave rise to hematite, which formed spherules in the rocks and, as these were worn away by the action of acidic water, they accumulated on the surface.

In relation to what was previously specified, and based on [9, 10], it is interesting to say that the soil of Meridiani Planum is composed of fine grains of basalt sand, in addition to a surface of spherules, with a high content of hematite, and other granules. According to [9], the erosion by action of the wind is visible, as well as small impacts of craters and layers of sedimentary rocks, finely laminated, rich in sulfides [10] and sulphated salts. Regarding cross lamination, it is known that on small scales it provides evidence that liquid water flowed through the study area. These rocks were probably a mixture of siliciclastic and chemical sediments formed in an episode of shallow water flooding followed by evaporation, exposure, and settling, similar to what occurs in a salt-marsh on Earth [11].

## 3. Material and methods

For this study, a total of 49 control points were selected, using planetocentric coordinates (**Table 1**), in the study area (**Figure 2**) based on the coverage of images downloaded from the Pilot tool [12], and captured by the Viking Orbiter between 1976 and 1980. The purpose, of using images from the Viking Orbiter, is none other than to check whether water and chlorophyll pigment-producing organisms existed on the surface of Mars in that period.

In total, 942 images were obtained, all in the planetocentric coordinate system, and whose geographic coordinates, which make up the area from which the control points have been established, have values, relative to the upper left corner of the study area, of 10° latitude and − 15° longitude, while those corresponding to the lower right corner are −10° latitude and 15° longitude.

In order to eliminate the possible atmospheric effects present in the information of each pixel of the downloaded images, the procedure patented by the main author of this work was used [13]. This procedure uses RGB coded bands

| Control point n° | Latitude (°) | Longitude (°) | Control point n° | Latitude (°) | Longitude (°) |
|---|---|---|---|---|---|
| **1** | 10 | −15 | **26** | 0 | 5 |
| **2** | 10 | −10 | **27** | 0 | 10 |
| **3** | 10 | −5 | **28** | 0 | 15 |
| **4** | 10 | 0 | **29** | −3 | −15 |
| **5** | 10 | 5 | **30** | −3 | −10 |
| **6** | 10 | 10 | **31** | −3 | −5 |
| **7** | 10 | 15 | **32** | −3 | 0 |
| **8** | 7 | −15 | **33** | −3 | 5 |
| **9** | 7 | −10 | **34** | −3 | 10 |
| **10** | 7 | −5 | **35** | −3 | 15 |
| **11** | 7 | 0 | **36** | −7 | −15 |
| **12** | 7 | 5 | **37** | −7 | −10 |
| **13** | 7 | 10 | **38** | −7 | −5 |
| **14** | 7 | 15 | **39** | −7 | 0 |
| **15** | 3 | −15 | **40** | −7 | 5 |
| **16** | 3 | −10 | **41** | −7 | 10 |
| **17** | 3 | −5 | **42** | −7 | 15 |
| **18** | 3 | 0 | **43** | −10 | −15 |
| **19** | 3 | 5 | **44** | −10 | −10 |
| **20** | 3 | 10 | **45** | −10 | −5 |
| **21** | 3 | 15 | **46** | −10 | 0 |
| **22** | 0 | −15 | **47** | −10 | 5 |
| **23** | 0 | −10 | **48** | −10 | 10 |
| **24** | 0 | −5 | **49** | −10 | 15 |
| **25** | 0 | 0 | | | |

**Table 1.**
*Control points on Mars surface where spectral signature has been obtained.*

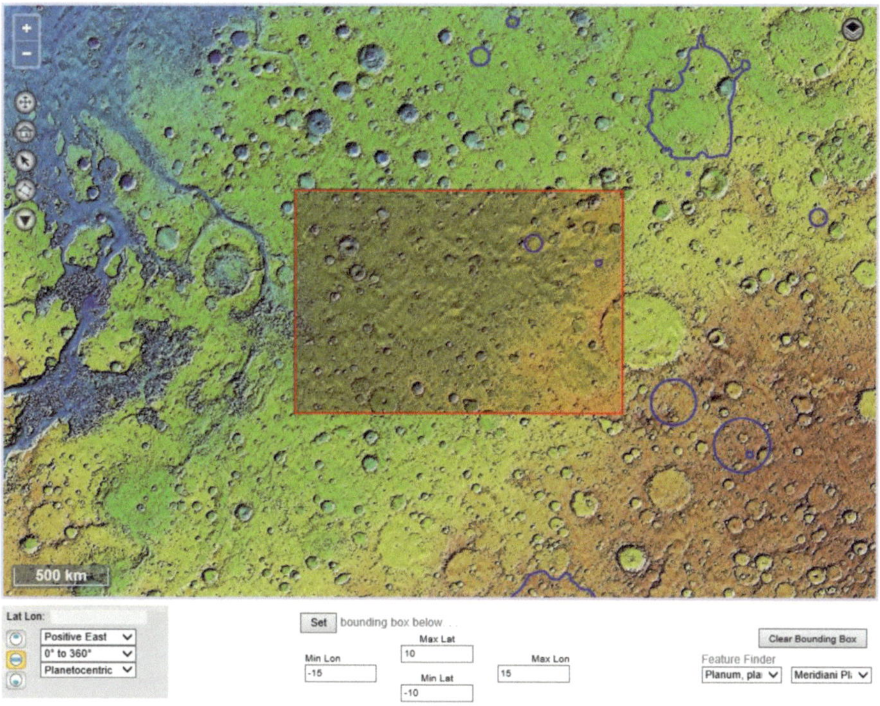


**Figure 2.**
*Location of the study area.*

(Red-Green-Red), which have a similar effect to that obtained when a satellite image is corrected for bias and field. The resulting images will be displayed in three colors (magenta, green and dark green), the latter being the one shown in those areas whose relative moisture content is high compared to the rest of areas of same image. It should be emphasized that the coding of visible bands, by means of the patented procedure obtained by [13], results in a series of spectral bands that provide more information than those corresponding only to the visible spectrum, since it allows to extract, from each pixel, the information found in the region from ultraviolet to near infrared.

In order to find algorithms capable of detecting both the moisture content and the existence of chlorophyll pigment-producing organisms on Mars surface, the spectral signatures of the 49 control points specified above have been obtained, each of which was compared with the spectral signatures encoded (with range of wavelengths in nm) of different types of water, as well as different types of ice, existing in certain areas of the Earth (**Table 2**), which are presented in **Figures 3** and **4**.

In the same way, a massive search for plant pigments was carried out at the pixel level, using the Plant Pigment Ratio (PPR) [14], to later relate it to the Coded Normalized Vegetation Index (CNDVI) obtained by [13]. Equations of the PPR and CNDVI indices are shown in Eqs. (1) and (2) respectively.

$$PPR\ (mol \cdot mol^{-1} of\ Clhorophill) = \frac{(Blue - Green)}{(Blue + Green)} \tag{1}$$

$$CNDVI = \frac{(Red\ cod - Green\ cod)}{(Red\ cod + Green\ cod)}\ (r = 0,85; R^2 = 0,91) \tag{2}$$

In another vein, a simulation of the transported soil was carried out, as a consequence of erosive processes, in the study area. To carry out this analysis, a total of 82

| | |
|---|---|
| **Water type 1** | Mean spectral signature of water on Earth |
| **Freshwater** | Mean spectral signature of freshwater on Earth |
| **Saltwater** | Mean spectral signature of saltwater on Earth |
| **Acid water 1** | Mean spectral signature of the acidic water belonging to the South Minas de Riotinto reservoir (Huelva, Spain) |
| **Acid water 2** | Mean spectral signature of the acidic water belonging to the North Minas de Riotinto reservoir (Huelva, Spain) |
| **Turbid water** | Mean spectral signature of turbid water existing in the Guadalquivir River (South of the Iberian Peninsula, Spain) |
| **Frozen water (type 1)** | Mean spectral signature of water ice on Earth |
| **Frozen water (type 2)** | Mean spectral signature of the Sierra Nevada water ice (Granada, Spain) |

**Table 2.**
*Different types of both water and ice, existing in the Earth, taken as reference.*

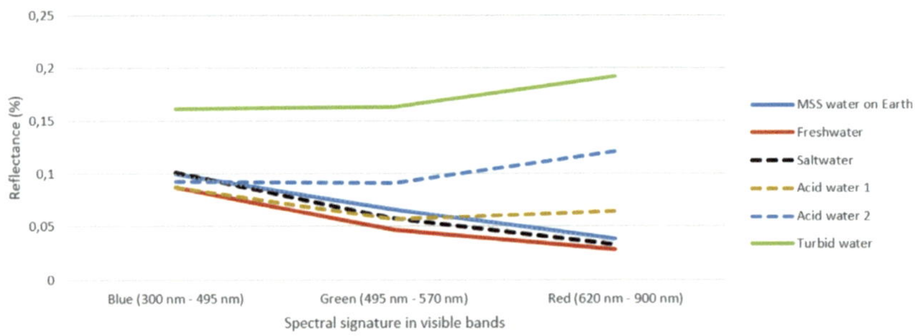


**Figure 3.**
*Spectral signatures of different types of water in the visible spectrum (MSS water on Earth is the Mean Spectral Signature of water on Earth).*

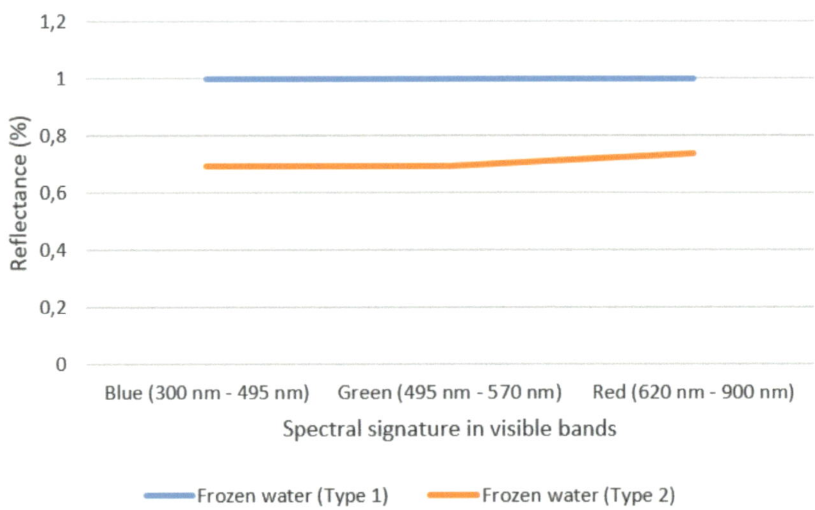


**Figure 4.**
*Spectral signatures of two types of frozen water in the visible spectrum.*

control points were established, in the form of a rectangular mesh, in the study area, although taking the area covered by Opportunity as a reference.

As a consequence of not having initial topographic values of the soil of Mars, nor of how they have varied over time, the total transported soil is calculated using the equation proposed by [11] for salt-marshes. It must be taken into account that the small diameter of the pebbles existing in the study area, which have been measured from the images captured by Opportunity using ImageJ software, are similar to the diameter of those existing in salt-marshes on Earth with a very slow water course velocity (such as that existing at Tinto or Odiel salt-marshes, or even at both Isla Cristina or Doñana salt-marshes). Therefore, the model used is shown in Eq. (3):

$$y = 0,032 - 0,133 \cdot SD - 1,164 \cdot T \; (r = 0,973; R^2 = 0,941) \qquad (3)$$

Where "y" is the prediction of volume of eroded soil in $hm^3$, "SD" is the equivalent area ($km^2$) where the eroded volume has occurred, and "T" is the height counted from the lowest level of the bed depth, in each control point, in study area (positive values represent sediment deposition, while negative values represent soil erosion).

In order to be able to calculate, with sufficient precision and through iterative processes [13], the volume of eroded soil, the Digital Terrain Model (DTM) of the Mars Orbiter Laser Altimeter (MOLA) has been used.

In relation to the above, a model was obtained, from which, to calculate the water flow velocity that existed in Meridiani Planum based on the pebbles diameter found in the Opportunity images. This model was compared with the one obtained by the second and third author of this chapter in a zone of slow water flow velocity of the Chaqui river in Bucay (Ecuador) and in a stream located in Bonita Creek (CA, USA) respectively.

## 4. Results and discussion

### 4.1 Analysis of spectral signatures

After comparing the spectral signatures of each of the 49 control points (**Table 1**) established in the study area, with the spectral signatures encoded (with range of wavelengths in nm) of the Earth (**Figures 3** and **4**), coincidences have been found in three of them (**Figure 5**), and more specifically with those corresponding to acid water 2, turbid water and frozen water (type 2), although only in 39 control points (**Table 3**).

On the other hand, and as a consequence of the existence of perchlorate salt (it enables the existence of acid water, in addition to facilitating the water evaporation at temperatures below 0°C) on Mars surface [15], a new algorithm, called Moisture Soil Index on Mars (MSIM), has been obtained, with which it is possible to obtain the humidity percentage in Martian soil. The expression of the MSIM is shown in Eq. (4):

$$MSIM\,(\%) = 10 \cdot \left[1 - \left(\frac{Blue\ band}{Red\ band}\right)^{1,5}\right] \; (r = 0,71; R^2 = 0,735) \qquad (4)$$

After comparing the MSIM with the WEH moisture index (it corresponds to the amount of hydrogen molecules in the water) established by [16] (**Figure 6**) it is observed that, although there is no relationship between both indices, it is true that a high WEH corresponds to high MSIM and vice versa.

In order to facilitate the use of this new index, an estimate of it can be obtained from the planetocentric coordinates of Mars, as shown in Eq. (5).

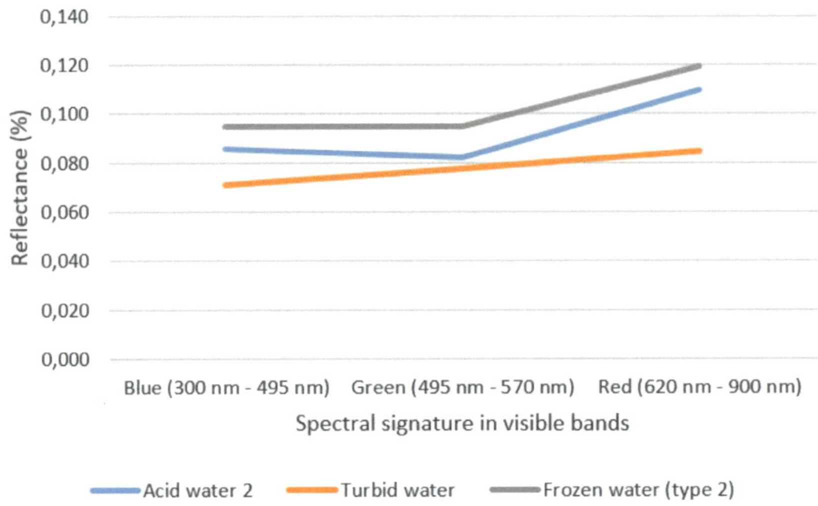


**Figure 5.**
*Spectral signatures found on Mars.*

| | Acid water 2 | | | | | | Turbid water | | | Frozen water (type 2) | |
|---|---|---|---|---|---|---|---|---|---|---|---|
| **C.P.** | **Lat** | **Long** | **C.P.** | **Lat** | **Long** | **C.P.** | **Lat** | **Long** | **C.P.** | **Lat** | **Long** |
| **1** | 10° | −15° | **17** | 3° | −5° | **11** | 7° | 0° | **18** | 3° | 0° |
| **2** | 10° | −10° | **21** | 3° | 15° | **19** | 3° | 5° | | | |
| **3** | 10° | −5° | **23** | 0° | −10° | **20** | 3° | 10° | | | |
| **4** | 10° | 0° | **24** | 0° | −5° | **22** | 0° | −15° | | | |
| **5** | 10° | 5° | **27** | 0° | 10° | **26** | 0° | 5° | | | |
| **6** | 10° | 10° | **28** | 0° | 15° | **31** | −3° | −5° | | | |
| **7** | 10° | 15° | **29** | −3° | −15° | **36** | −7° | −15° | | | |
| **8** | 7° | −15° | **30** | −3° | −10° | **41** | −7° | 10° | | | |
| **9** | 7° | −10° | **34** | −3° | 10° | **43** | −10° | −15° | | | |
| **10** | 7° | −5° | **35** | −3° | 15° | **44** | −10° | −10° | | | |
| **13** | 7° | 10° | **37** | −7° | −10° | **47** | −10° | 5° | | | |
| **14** | 7° | 15° | **45** | −10° | −5° | | | | | | |
| **15** | 3° | −15° | **49** | −10 | 15° | | | | | | |
| **16** | 3° | −10° | | | | | | | | | |

**Table 3.**
*Planetocentric coordinates of Mars with coincidences in spectral signature with Earth (C.P. = Control Point).*

$$MSIM\ (\%) = -0,32 + 0,014 \cdot L + 0,063 \cdot Lat - 0,033 \cdot Lon\ \left(r = 0,71; R^2 = 0,47\right) \tag{5}$$

Knowing that:

L = Luminance of each pixel = Red + Green + (Blue/3). It is a dimensionless index.

Lat = Latitude in decimal degrees.

Lon = Longitude in decimal degrees.

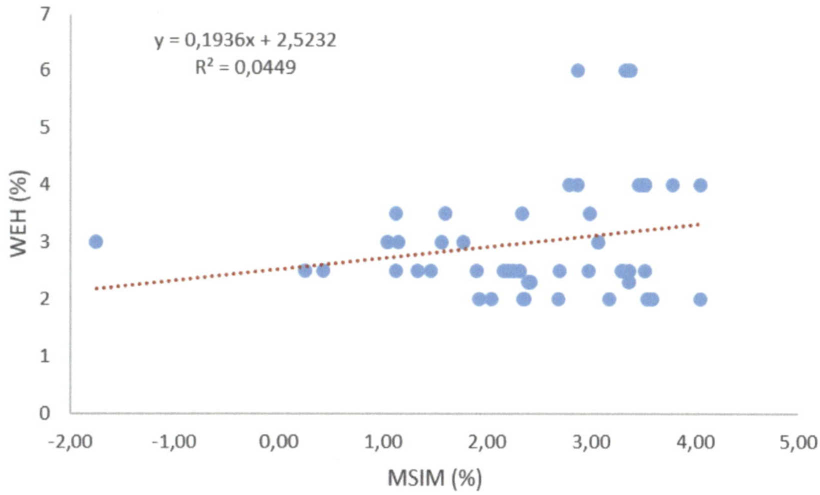


**Figure 6.**
*Comparison between WEH [16] and MSIM obtained for Mars in the study area.*

### 4.2 Analysis of moisture sources from Opportunity images

As is well known, according to [17], soil moisture plays a fundamental role both in the hydrological cycle and in land-atmosphere interactions. However, there are several studies in which the importance of the soil has been analyzed in various scientific areas, such as in climate simulations and meteorological prediction [18], in precipitation and runoff models [19] or in evapotranspiration [20] among others.

In this sense, and with regard to Eagle Crater, [21] specified that the outcrop, shown in the upper part of **Figure 7**, was wet by salt water, in addition to containing high concentrations of hematite and jarosite, minerals that form from the alteration of rocks in the presence of acidic water.

At this point, it is very important to mention that, according to [23], depending on the texture and structure of the soils, these can contain a certain amount of moisture that is normally generated as a result of a rainfall event [24], that saturates the soil causing a portion to drain towards the interior due to the gravity force.

Regarding the Endurance crater, Opportunity made a total of seven holes using the rock abrasion tool. Likewise, the rover made combined mosaic images like the one shown in **Figure 8** on the left. Evidence of fine-grained red hematite was observed around the drill area [21].

It is convenient to emphasize that the dark green area, selected with the red arrow in **Figure 8** right, contains iron oxide compounds, which, on the surface of Mars, have a color similar to that of hematite on Earth (gray).

In another vein, according to [26], the studied area by Opportunity have sandstones rich in sulfates, which indicates the existence of past erosion by the action of both the wind and by a slow water flow as a result of the possible formation of ephemeral (and shallow) lakes.

In **Figure 9** right it can be seen that the dark green coloration, which indicates the presence of moisture [13], is found in the predominant stratum on the right. This stratum is formed by a lumpy rock called Wopmay located on the inner of Endurance slopes. It is believed that this type of rock is the consequence of an alteration due to the presence of water [27].

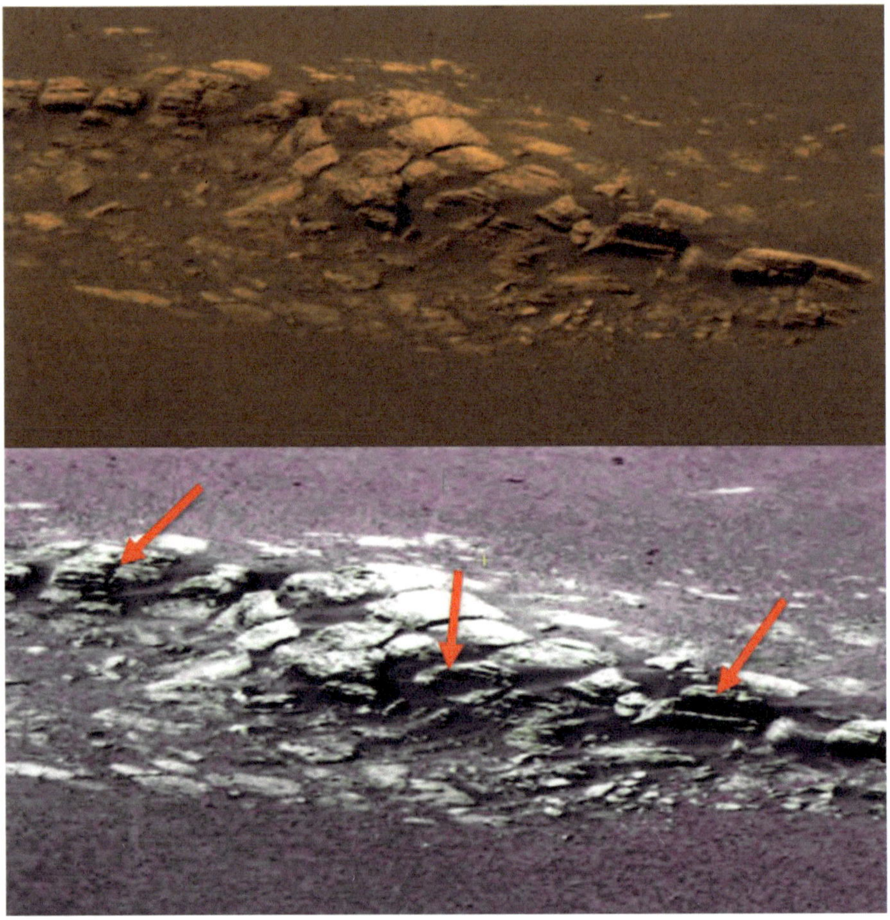

**Figure 7.**
*Image taken by Opportunity (Sol 4) at Eagle Crater [22]. The lower image shows the wet areas (red arrows) after subjecting the upper image to the process patented by [13].*

### 4.3 Model to obtain water flow velocity based on pebbles diameter

To obtain this model, both in the Chaqui River (Bucay, Ecuador) and in a stream located in Bonita Creek (California, USA), the water velocity was calculated at a series of points established in field (15 in Ecuador and 10 in the USA), to later measure the average pebbles diameter at those same points. Subsequently, with the use of genetically modified algorithms, although taking into account the gravity of Mars (3,7 m/s$^2$), the water flow velocity that should exist on Mars was calculated to have pebbles of the same diameter as those measured in the specified sampling points. For this process, the DTM obtained by MOLA was also used. The result was that the water flow velocity on Mars having to be in a relationship as shown in Eq. (6):

$$V_{on\ Mars} = \frac{e^{D_{on\ Earth}}}{3}\ \left(r = 0,91; R^2 = 0,99\right) \tag{6}$$

Once the water velocity on Mars was obtained, it was represented against the diameter, obtaining the model presented in **Figure 10**, and from which it is possible to calculate the water velocity in any area of Mars as long as there is evidence that it was covered by water. In Eq. (7) the model presented in **Figure 10** is appropriately specified.

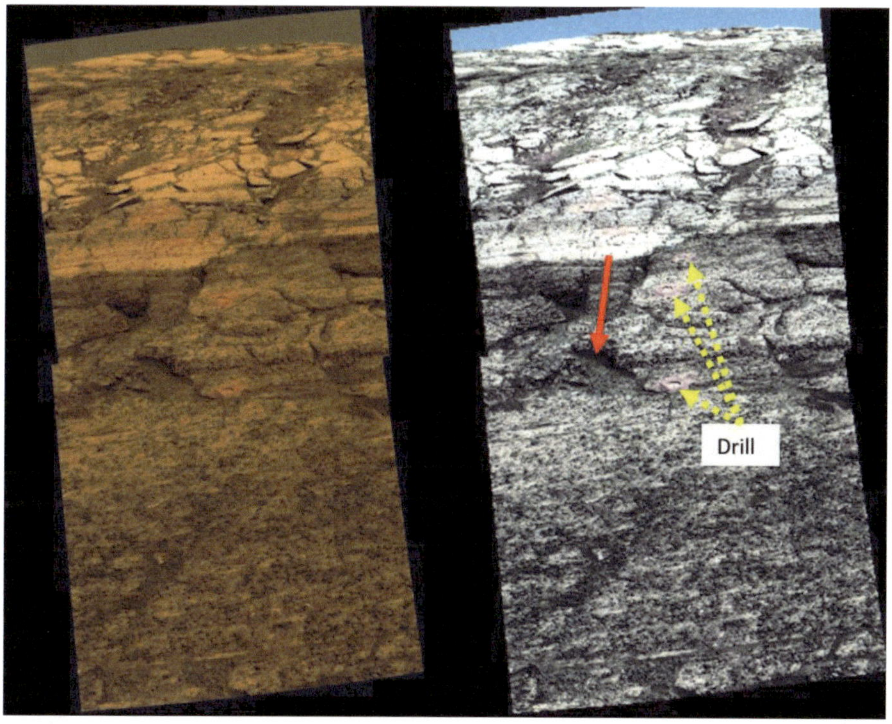


**Figure 8.**
*Image taken by Opportunity (Sol 173) at Endurance Crater [25]. The right image shows the wet areas (red arrow) after subjecting the upper image to the process patented by [13].*

$$V_{on\ Mars} = 0,4961 \cdot D_{pebbles\ on\ Mars} + 0,3025\ \left(r = 0,98; R^2 = 0,99\right) \tag{7}$$

Authors want to record that the model shown in Eq. (7) has been used by two studies, pending publication, in the areas of the Spirit and Curiosity rovers respectively, obtaining results that are fully consistent with the reality of Mars.

Later, with the use of the free software ImageJ (**Figure 11**), the pebbles diameter in the study area was measured. In **Figure 11** is shown a measure that no corresponding to a pebbles only to show, clearly, the procedure used. Pebbles diameters measured in study area fit perfectly (see **Figure 10**) with Eq. (7).

### 4.4 Prediction of erosion in Meridiani Planum

As can be seen in **Figure 10**, as well as in Eq. (7), the water velocities obtained on Mars are in the same velocities range as those obtained in a salt-marsh, for this reason the model shown in Eq. (3) and proposed by [11] has been used.

The transport of sediments in study area has been calculated through an evaluation to pixel level (**Figure 12**) using the DTM obtained by MOLA and the patented procedure by [13]. In this figure, sediment transport is simulated based on climatic conditions [28] on Mars and a surface shape factor dependent of algorithm used by [13].

Throughout its history, and according to [24], the surface of Mars has had a very active hydrological cycle, which has given rise to a network of valleys formed, mainly, by the surface flow erosion, leaving, as secondary factor, erosion due to groundwater seepage. It is precisely this fact that makes it possible for the depth profile at the control points 12, 15 and 77 to reach the values −1437,47 m, −1676,72 m and − 2873,03 m respectively.

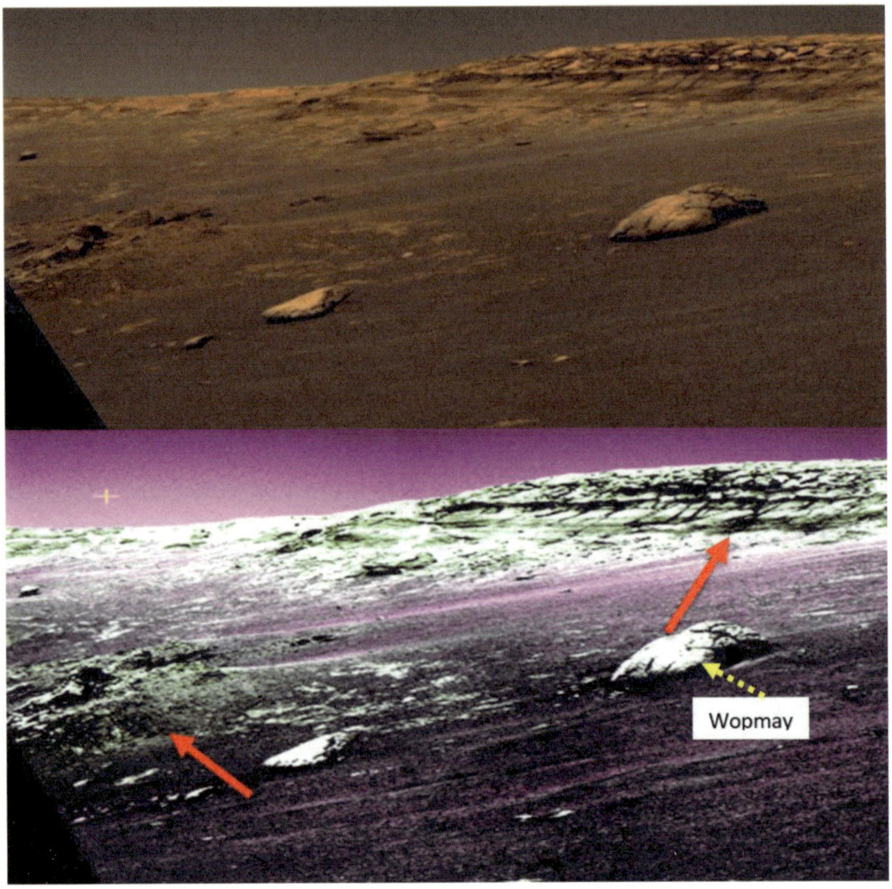


**Figure 9.**
*Image taken by Opportunity (Sol 248) at Endurance Crater [27]. The right image shows the wet areas (red arrows) after subjecting the upper image to the process patented by [13].*

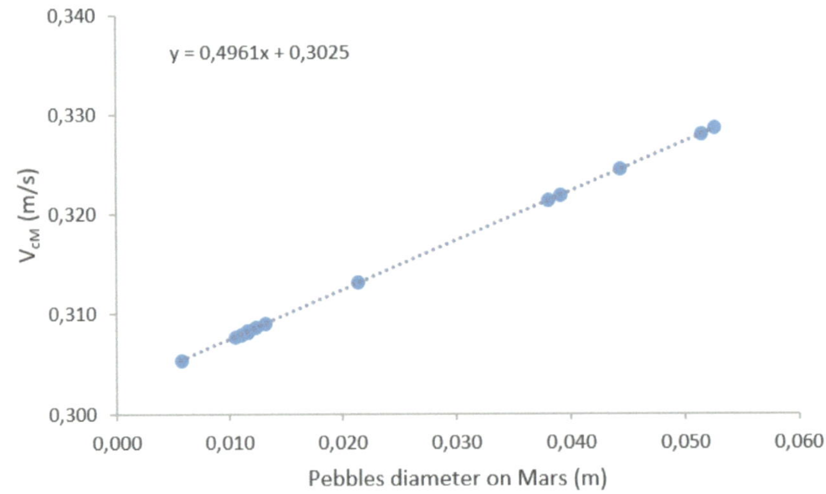


**Figure 10.**
*Relationship between the water velocity (m/s) and the pebbles diameter (m) on Mars surface.*

**Figure 11.**
*Procedure used to measure diameter (length measurements are in pixels).*

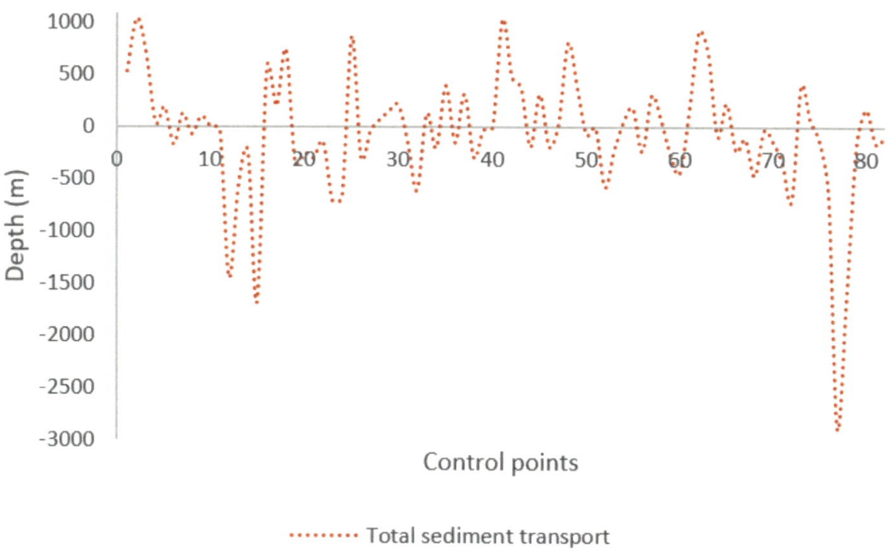


**Figure 12.**
*Variation of the depth in terms of sediment transport in study area (positive values = soil deposition; negative values = soil erosion).*

### 4.5 Possible existence of chlorophyll pigment-producing organisms on Mars surface

After the analysis carried out at the pixel level, in order to find some trace of vegetal life, it has been obtained that the PPR index [14] has varied between $-0,10$ and $0,09$ mol $\cdot$ $mol^{-1}$ of chlorophyll (**Figure 13**). It must be specified that on Earth, values between 0 and 0,1 mol $\cdot$ $mol^{-1}$ chlorophyll (both inclusive) correspond to Neoxanthin.

According to [29], Neoxanthin is a type of carotenoid and xanthophyll existing in some types of plants, and whose function is to protect the plant against photo-oxidative stress.

It must be taken into account that, on Earth, the production of Neoxanthin by vegetables remains constant at temperatures of up to $-20°C$, easily achievable on Mars, while the production of Carotenoids and Beta-Carotenoids decreases at temperatures below 0°C. For this reason, there is a certain probability that organisms that synthesize, mainly, Neoxanthin may exist on Mars surface.

After the bibliographic review carried out, and according to [30], one of the organisms that synthesizes Neoxanthin, and that may possibly be present on the Mars surface, are microalgae with a protected cell wall. These organisms can survive in extreme conditions as long as they are adapted to the environment.

It is known that microalgae proliferate on Earth in conditions of contamination of the aqueous environment, which can be found on Mars due to the existence of perchlorate salt. In the same way, and after the analysis of the spectral signatures, acidic water and turbid water have been found, so the existence of microalgae is reinforced.

Given the possible existence of microalgae on Mars, the new CNDVI (Eq. (2)) was obtained, whose relationship with the PPR index is presented in **Figure 14**. As can be seen in **Figure 14**, relationship between both variables is quite good, so the idea of the possible existence of microalgae, or Neoxanthin-producing organisms, on Mars becomes more consistent.

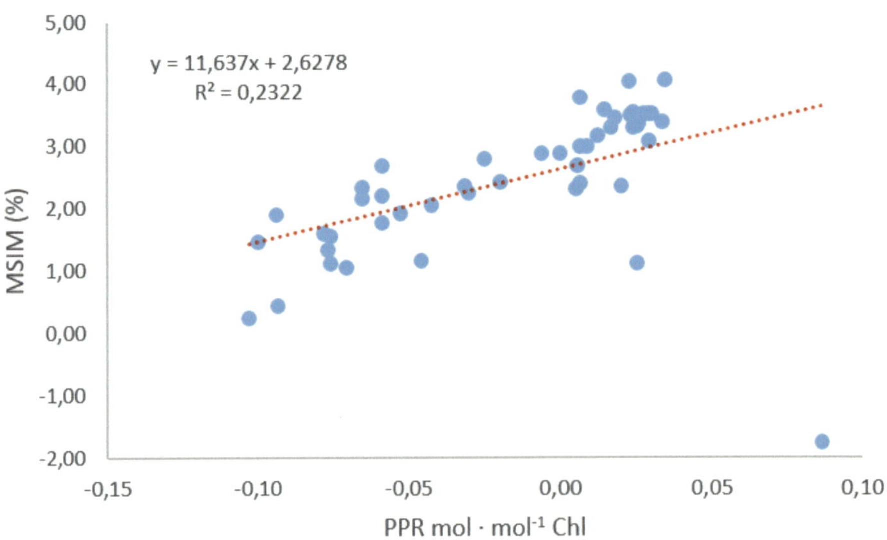


**Figure 13.**
*Relationship between the MSIM and the PPR index in study area.*

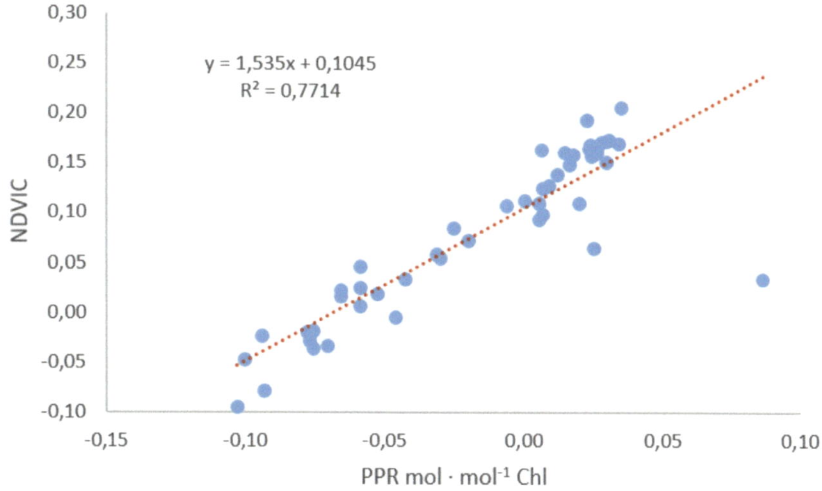


**Figure 14.**
*Relationship between the CNDVI and the PPR index in study area.*

## 5. Conclusions

In this study, different novel aspects related to Meridiani Planum, as well as Opportunity, have been shown.

Logically, despite the fact that the results obtained are encouraging in many aspects, other hypotheses must be taken into account in order to complement not only this research work, but also all those that are carried out from this study.

However, we must pay attention to the well-known seasonal process of water mobility on Mars as a result of perchlorate salts, since a phenomenon similar to FVAT [31] can exist, due to which the transit of water and salt of perchlorate towards the surface is possible from certain environmental conditions of relative humidity, pressure, temperature, as well as a determined concentration level of perchlorate salt in the water, be it acidic or turbid.

During the present work, the presence of aqueous minerals has been corroborated, although it is true that the samples collected by Opportunity, as well as other rovers that are in operation on Mars, have sufficient reliability. In this regard, this study shows that the data obtained (from the 942 images analyzed) by the Viking Orbiter have the same reliability, especially as a result of the fact that they have made it possible to corroborate aspects, phenomena and situations that occurred on Mars, and that Opportunity, Spirit and Curiosity had demonstrated. A special mention should be made of Perseverance, whose data will allow us to advance our knowledge of the red planet.

Authors want to highlight, given that they have carried out the same study in the areas covered by Spirit and Curiosity, that the results obtained are similar, although in the area covered by Curiosity there is evidence of a lower amount of PPR. On the other hand, in the Gusev crater, area where Spirit is located, a spectral signature corresponding to saltwater has been found.

Finally, it must be emphasized the importance of new technologies, patented processes, as well as the development of new advanced calculation tools capable of jointly simulating all possible situations that may occur on any rocky planet or moon existing, since in this way the advancement of knowledge in those worlds will be more efficient.

## Conflict of interest

Authors declare no conflict of interest.


## Author details

Emilio Ramírez-Juidías[1]*, Katherine Villavicencio-Valero[2] and Arthur Borja[3]

1 University of Seville, Seville, Spain

2 International Research School of Planetary Sciences (U.G. D'Annunzio), Pescara, Italy

3 Bayer School, USA

*Address all correspondence to: erjuidias@us.es

## References

[1] Greenwood RC, Burbine TH, Franchi IA. Linking asteroids and meteorites to the primordial planetesimal population. Geochimica et Cosmochimica Acta. 2020;277: 377-406. DOI: 10.1016/j. gca.2020.02.004

[2] Gaffey MJ. The early solar system. In: Whittet DCB, editors. Planetary and Interstellar Processes Relevant to the Origins of Life. Springer, Dordrecht; 1997. p. 185-203. DOI: 10.1007/978-94-015-8907-9_10

[3] Rauchfuss H, Mitchell TN. The cosmos, the solar system and the primeval Earth. In: Chemical Evolution and the Origin of Life. Springer, Berlin, Heidelberg; 2009. p. 17-41. DOI: 10.1007/978-3-540-78823-2_2

[4] Manske L, Marchi S, Plesa AC, Wünnemann K. Impact melting upon basin formation on early Mars. Icarus. 2021;357:1-13. DOI: 10.1016/j. icarus.2020.114128

[5] Marchi S, Walker RJ, Canup RM. A compositionally heterogeneous Martian mantle due to late accretion. Science Advances. 2020;6:1-7. DOI: 10.1126/ sciadv.aay2338

[6] Catling DC, Zahnle KJ. The planetary air leak. Scientific American. 2009;300 (5):36-43. DOI: http://www.jstor.org/ stable/26001341

[7] USGS. Planetary names [Internet]. 2021. Available from: https:// planetarynames.wr.usgs.gov/images/ mola_regional.pdf [Accessed: 2021-02-05]

[8] NASA/JPL-Caltech/Arizona State University. Mars exploration program: Meridiani planum [Internet]. 2012. Available from: https://mars.nasa.gov/ resources/5260/meridiani-planum/ [Accessed: 2021-02-05]

[9] Hynek BM, Di Achille G. Geologic map of Meridiani Planum, Mars. US Geological Survey Scientific Investigations Map [Internet]. 2017; 3356:1-13. Available from: http://citesee rx.ist.psu.edu/viewdoc/download?doi= 10.1.1.1071.8324&rep=rep1&type=pdf [Accessed: 2021-02-05]

[10] McCollom M, Hynek B. Geochemical data indicate highly similar sediment compositions for the Grasberg and Burns formations on Meridiani Planum, Mars. Earth and Planetary Science Letters. 2021;557:1-13. DOI: 10.1016/j.epsl.2020.116729

[11] Ramírez-Juidías E, Víquez-Urraco F, Noguero-Hernández D. Sedimentary processes in the Isla Cristina salt-marshes: geomorphological changes of landscape. Ocean & Coastal Management. 2017;143:148-153. DOI: 10.1016/j.ocecoaman.2016.11.007

[12] USGS, NASA. Planetary Image Locator Tool [Internet]. 2008. Available from: https://pilot.wr.usgs.gov/ [Accessed: 2021-01-10]

[13] Ramírez-Juidías E, Pozo-Morales L, Galán-Ortiz L. Procedure for ontaining a remote sensed image from a photograph. Patent n° ES2537783B2 (2015-09-29 publication of the patent concession). International Patent n° WO2014198974A1. Universidad de Sevilla. 2015. Available from: http:// consultas2.oepm.es/InvenesWeb/deta lle?referencia=P201300573

[14] Metternicht G. Vegetation indices derived from high-resolution airbone videography for precision crop management. International Journal of Remote Sensing. 2003;24(14):2855-2877. DOI: 10.1080/01431160210163074

[15] Moskowitz C. The Mars Resonnaissance Orbiter found evidence that flowing water causes suspicious

dark streaks on the Red Planet. Scientific American [Internet]. 2015. Available from: https://www.scientificamerican.com/article/water-flows-on-mars-today-nasa-announces/ [Accessed: 2021-02-15]

[16] Wilson JT, Eke VR, Massey RJ, Elphic RC, Feldman WC, Maurice S, Teodoro LFA. Equatorial locations of water on Mars: improved resolutions maps based on Mars Odyssey Neutron Spectrometer data. Icarus. 2018;299: 148-160. DOI: 10.1016/j.icarus.2017.07.028

[17] Mattar C, Sobrino JA, Wigneron JP, Jiménez-Muñoz JC, Kerr Y. Estimación de la humedad del suelo a partir de índices de vegetación y microondas pasivas. Revista de Teledetección. 2011; 36:62-72. DOI: http://www.aet.org.es/?q=revista36-8

[18] Pinnington E, Amezcua J, Cooper E, Dadson S, Ellis R, Peng J, Robinson E, Morrison R, Osborne S, Quaife T. Improving soil moisture prediction of a high-resolution land surface model by parameterizing pedotransfer functions through assimilation of SMAP satellite data. Hydrology and Earth System Science. 2021;25:1617-1641. DOI: 10.5194/hess-25-1617-2021

[19] Cheruy F, Ducharne A, Hourdin F, Musat I, Vignon E, Gastineau G, Bastrikov V, Vuichard N, Diallo B, Dufresne JL, Ghattas J, Grandpeix JY, Idelkadi A, Mellul L, Maignan F, Ménégoz M, Ottlé C, Peylin P, Servonnat J, Wang F, Zhao Y. Improved near-surface continental climate in IPSL-CM6A-LR by combined evolutions of atmospheric and land surface physics. Journal of Advances in Modelling Earth Systems. 2020;12(10):1-33. DOI: 10.1029/2019MS002005

[20] Chen JM, Liu J. Evolution of evapotranspiration models using thermal and shortwave remote sensing data. Remote Sensing of Environment. 2020;237:1-20. DOI: 10.1016/j.rse.2019.111594

[21] NASA. The mars Exploration Rovers: Spirit and Opportunity. [Internet]. 2013. Available from: https://mars.nasa.gov/files/resources/MER10-YearAnniversaryLithograph.pdf [Accessed: 2021-02-16]

[22] NASA/JPL/Cornell. Eagle Crater (Mars Exploration Rover Opportunity). [Internet]. 2004. Available from: https://mars.nasa.gov/mer/gallery/press/opportunity/20040506a/02-SS-02-OpportunityLedge-B101R1_br.jpg [Accessed: 2021-02-18]

[23] Zotarelli L, Dukes MD, Morgan KT. Interpretation of soil moisture content to determine soil field capacity and avoid over-irrigating sandy soils using soil moisture sensors. EDIS. 2010;2:1-4. DOI: http://edis.ifas.ufl.edu/ae460

[24] Seybold HJ, Kite E, Kirchner JW. The role of surface water in the geometry of Mars' valley networks and its climatic implications. [Internet]. 2017. Available from: https://arxiv.org/abs/1709.09834 [Accessed: 2021-02-16]

[25] NASA/JPL/Cornell. Endurance Crater (Mars Exploration Rover Opportunity). [Internet]. 2004. Available from: https://photojournal.jpl.nasa.gov/catalog/PIA06727 [Accessed: 2021-02-18]

[26] Arvidson RE, Ashley JW, Bell III JF, Chojnacki M, Cohen J, Economou TE, Farrand WH, Fergason R, Fleischer I, Geissler P, Gellert R, Golombek MP, Grotzinger JP, Guinness EA, Haberle RM, Herkenhoff KE, Herman JA, Iagnemma KD, Jolliff BL, Johnson JR, Klingelhöfer G, Knoll AH, Knudson AT, Li R, McLennan SM, Mittlefehldt DW, Morris RV, Parker TJ, Rice MS, Schröder C, Soderblom LA, Squyres SW, Sullivan RJ, Wolff MJ. Opportunity Mars Rover mission: overview and selected results from

Purgatory ripple to traverses Endeavour crater. Journal of Geophysical Research. 2011;116:1-33. DOI: 10.1029/2010JE003746

[27] NASA/JPL/Cornell. Wopmay rock in Endurance Crater (Mars Exploration Rover Opportunity). [Internet]. 2004. Available from: https://photojournal.jpl.nasa.gov/catalog/PIA06920 [Accessed: 2021-02-18]

[28] CSIC-INTA. Rover Environmental Monitoring Station. [Internet]. 2009. Available from: http://cab.inta-csic.es/rems/es/atmosfera-de-marte/ [Accessed:2021-01-06]

[29] Dall'Osto L, Cazzaniga S, North H, Marion-Poll A, Bassi R. The Arabidopsis aba4-1 mutant reveals a specific function for neoxanthin in protection against photooxidative stress. The Plant Cell. 2007;19:1048-1064. DOI: 10.1105/tpc.106.049114

[30] García S, Pérez RM. Aspectos analíticos sobre la determinación de compuestos carotenoids en microalgas mediante cromatografía de líquidos con detector de diodos. [Internet]. 2012. Available from: https://www.osti.gov/etdeweb/servlets/purl/21542049 [Accessed: 2021-02-07]

[31] Ramírez-Juidías E, Yanes-Figueroa J. Estudio medioambiental del municipio de almadén mediante teledetección. Interciencia. 2017;42(3):157-162. DOI: https://www.interciencia.net/wp-content/uploads/2017/08/157-162-5834-RAMIREZ-42_3.pdf

Chapter 5

# Life on Mars: Clues, Evidence or Proof?

*Vincenzo Rizzo, Richard Armstrong, Hong Hua, Nicola Cantasano, Tommaso Nicolò and Giorgio Bianciardi*


## Abstract

The search for life on Mars is one of the main objectives of space missions. At "Pahrump Hills Field Site" (Gale Crater, Mojave target), inside the mudstones of the Murray lacustrine sequence, Curiosity rover found organic materials and lozenge shaped laths considered by NASA as pseudomorphic crystals. Besides it detected mineral assemblages suggesting both oxidizing (hematite) and reducing (magnetite) environments, as well as acidic (diagenetic and/or authigenic jarosite) and neutral (apatite) conditions, that might suggest bacterially mediated reactions. Our morphological and morphometrical investigations show that such diagenetic microstructures are unlikely to be lozenge shapes and, in addition to several converging features, they suggest the presence of remnants of complex algal-like biota, similar to terrestrial procaryotes and/or eukaryotes; possible microorganisms that, on the base of absolute dating criteria used by other scholars, lived on Mars about 2.12 +/−0.36 Ga ago.


**Keywords:** life on Mars, microalgae, microbialites, Mojave target, Gale Crater, Mars, Curiosity rover

## 1. Introduction

### 1.1 Mars planet: environment and geological history

As it is known, Mars is the fourth and the last of the inner and rocky planets of our Solar system. Beyond the "Red Planet", it extends the belt of asteroids just before the giant Jupiter and the other gaseous external planets of the system. It has a mass: 6.417 x $10^{23}$ Kg, a density of 3,940 g/cm$^3$, equatorial diameter of 6,792 Km and a mean temperature of −63°C, a distance from Sol of 228,000,000 km.

Mars orbits the Sun at an average distance of 230 million km and its revolution period is about 687 days; while his solar day is a little longer than ours: 24 hours, 37 minutes and 23 seconds. The Martian axial inclination is 25.19°, which is similar to that of Earth. Due to the discrete eccentricity of its orbit of 0.093, its distance from Earth to opposition can range between about 100 and about 56 million kilometers; only Mercury has a higher eccentricity in the Solar System. However, Mars used to follow a much more circular orbit: about 1.35 million years ago its eccentricity was equivalent to 0.002, which is much lower than the current Earth's. Mars has an eccentricity cycle of 96,000 Earth years compared to Earth's 100,000. Over the past 35,000 years, Martian orbit has become increasingly eccentric due to the gravitational influences of other planets, and the closest point between Earth and Mars will continue to decline over the future time.

The planet is enveloped by a thin atmosphere dominated by the gas carbon dioxide as the 95.3% of the whole. The other chemical elements with their respective rates are: nitrogen (2.7%), Argon (1.6%) and oxygen–carbon monoxide-water steam (0.4%). The weakness of the Martian atmosphere and the lack of a magnetic field do not allow any effective defence against ultraviolet radiations and solar winds.

The environmental and geological history of early Mars is mostly written in its rocks, in their composition and morphological/structural signatures. The first few billion years of Mars' geologic history records surface environments considerably different than the surface today, prompting a succession of coordinated surface and orbiter missions over the past two decades aimed ultimately at determining if Mars ever had an early biosphere. Water is the sine qua non for all life as we understand it and, therefore, past missions have sought environments where water was abundant and possibly long-lived [1, 2].

Mars was formed 4.6 billion years ago, with a history similar to the other three Terrestrial-type planets, i.e. as a result of the condensation of the solar nebula, most probably silicates. Due to the upper distance from the Sun from Earth, during the initial phase of formation in Mars' orbit there was a higher concentration of elements with low boiling points, such as chlorine, phosphorus and sulfur, probably driven away from the inner orbits by the strong solar wind of the young Sun [3].

During the first Martian Era, known as Noachian period, 4.1–3.7 Gys ago, its environment, that had formed, was moderately similar to the one on present Earth. Liquid water was widespread in a neutral environment, volcanic activity and heat flow more vigorous, and atmospheric pressure and temperature were higher than today. Morphological evidences are represented by river delta and river meanders, drainage networks and lakes; such morphologies are accompanied by the occurrence of consistent sedimentary, layered deposits. These conditions may have favoured the spread of life on the surface of Mars [4]. In this period the planet was subject to intense late bombardment, to which Earth was also a victim. In fact, about 60% of the surface has markers dating back to that era, particularly impact craters. The largest of these is located in the northern hemisphere and has a diameter of about 10,000 km, almost half the circumference of the planet. The formation of this crater is probably due to the impact with a big asteroid, which left a deep depression (the Boreal Basin), covering about 40% of the planet, brutally changing the history and environment of planet [5]. Eloquents morphological rest of such big impacts are extensive water flow formations in the Tharsis region; a region subject, towards the end of the era to a very active volcanism and flooded, by a large amount of water.

Slowly, in just over a billion and a half years, Mars went from a warm and humid phase characteristic of the Noachian to that of a cold and on-surface arid planet observable today. This transitional phase occurred during the Hesperian; a period characterized by continuation of intense volcanic activity (like those of Olympus Mons), deposition of evaporitic sedimentary sequences, and catastrophic floods that dug immense canals along the surface [6]. The continuous eruptions brought large amounts of sulphur dioxide and hydrogen sulfide to the surface, changing the large expanses of liquid water into small basins of high acidity water due to the sulphuric acid that formed. Although the disappearance of rivers and lakes is generally considered attributable towards the end of this era, recent dating on Gale Crater outcroppings open up the existence of water lake about billion years ago, during the Amazonian era [7].

One era, this last, from about 3 billion years ago to today, that is characterized by a poor period of meteor bombardment and by a continuation of cold and arid

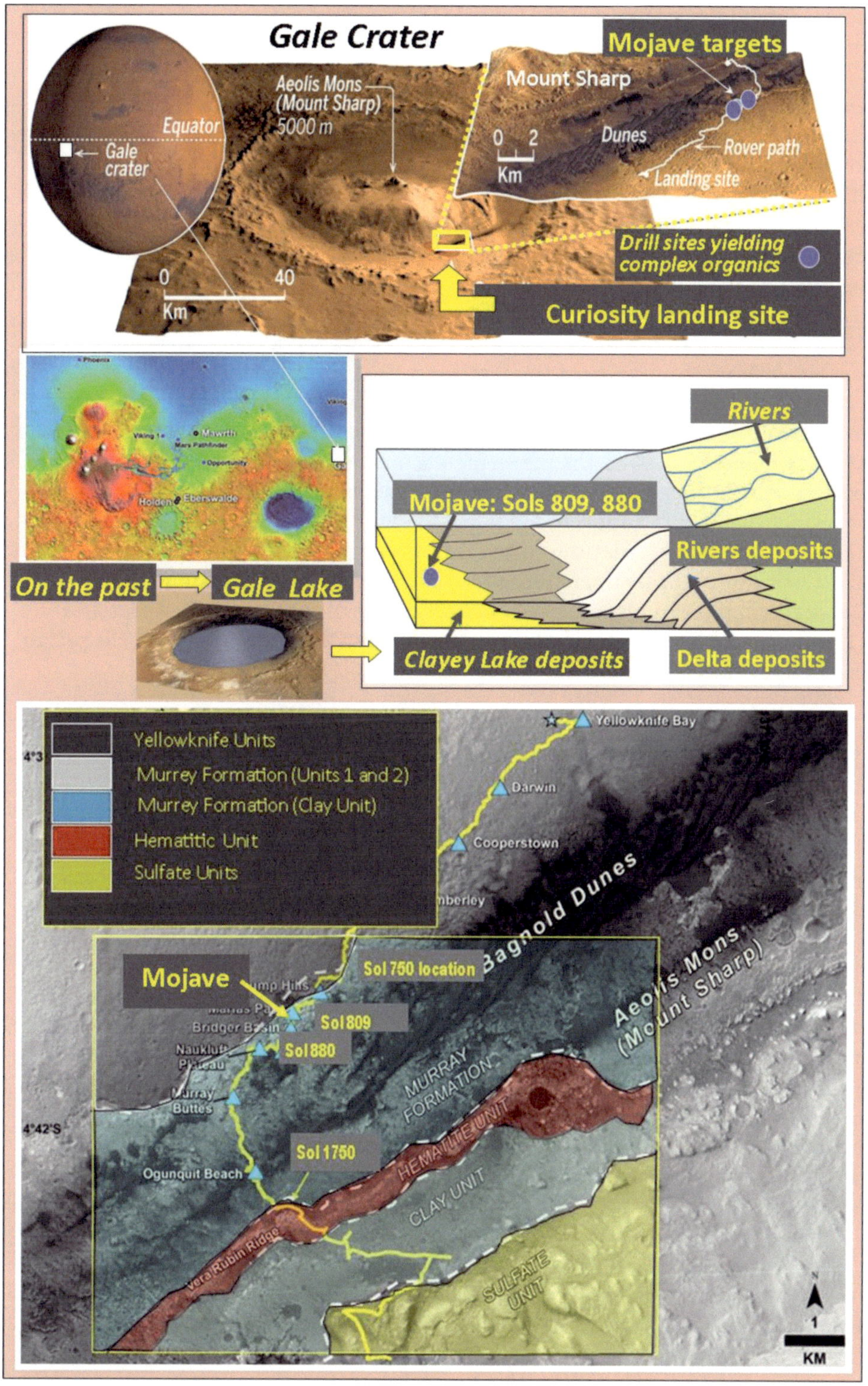


**Figure 1.**
*Pictures show the landing site of NASA Curiosity rover, at GaleCrater, located near the Martian Equator; a site where intertidal and lacustrine clayey deposit have been found, as well as organic molecules and microstructures resembling microbialtes and terrestrial algae have been found. Below: a geological sketching of the ancient lake deposits (from NASA reports, modified).*

climatic conditions, until today. Conditions producing surface aridity and deepening of water tables, whose existence has been proven by recent space missions.

In fact, the Martian poles are covered with water ice and permafrost layer extends to latitudes of about 60°, and large amounts of water are believed to be trapped under the thick Martian cryosphere [8]. The formation of Valles Marineris and its spill channels and sink holes, shows that during the early stages of Mars' history there was a large amount of liquid water. The presence of large amount of ground water and ice in the south pole of Mars was confirmed by the European Mars Express probe in January 2004 and by MARSIS radar near the Chryse Planitia region. Radar analyses conducted from 2012 to 2015 by Mars Express revealed presence of liquid salt water under the southern ice cap. In 2015, based on the MRO's monitoring, NASA announced evidence that liquid salt water flows on the surface of Mars in the form of small streams [9].

Recent data of spatial missions and in particular of NASA rovers instrumentation proved on Gale crater the existence of lacustrine and intertidal deltas deposits (**Figure 1**), presence of all the life elements, occurrence of complex organic molecules, sedimentary structures similar to terrestrial microbialites and the existence of an environment favorable to microbial life [1, 10, 11].

Stromatolite-like structures were also found by NASA rovers Opportunity, Spirit and Curiosity on the laminated Martian outcroppings of Meridiani Planum [12–18]. These findings have given strength to the hypothesis of stromatolites presence [19], that could be probably quite widespread from the Noachian to the Hesperian geological era of Martian life.

## 2. The search of life on Mars

The search for life on Mars, either in the present or in the past history of the "Red Planet", has been the main motivation behind research programs since the 1970s. The first images, highlighting the evidence for past liquid water on Mars, were carried out by orbital images from Mariner 9 [20]. Then, Mars Observer Camera and Mars Reconnaissance Orbiter [21] provided new images of past fluvial networks on its surface, long time ago.

In fact, the claim of this question stems from the early debates that arose from the Vikings' LR experiments until their recent re-analysis [84]. We can also recall as in 2004, the Phoenix space probe showed the existence of an ice-cemented ground in the northern plains of Mars combined with the presence of perchlorates normally used in terrestrial metabolic pathways of a large number of microorganisms [22–24]. So, perchlorates, found in the ground ice of Mars, is a putative biosignature resulting by a possible microbial activity on Mars [25] and could suggest a chemosynthetic activity carried out by bacteria on the planet surface, a short time ago [26].

Our knowledge about this topic has been increased considerably as a result of recent NASA missions, Opportunity, Spirit and Curiosity, located at Meridiani Planum, Gusev and Gale Craters, respectively. The Curiosity rover landed inside the Aeolis Crater, informally known as Gale Crater, on August 2012 with a complex set of scientific instruments (MSL), able to detect chemical and mineralogical soil composition, environmental data, and record panoramic and microscopic images of high accuracy, obtaining subsequently several thousands of images. Hence, the MSL scientists discovered a fluvial-lacustrine sequence and fine grained sedimentary rocks containing clay and hydrated minerals, deposited inside an almost neutral lake. These deposits were, then, subjected to two post-genetic more acidic phases, revealing that they contain the elements necessary for life, e.g., H, O, S, C, N, P, and also including Fe, Mg and Mn suitable to support a possible Martian biosphere based on chemolithoautotrophy [1]. In addition, life, if it existed, must have left visible traces of its activity and presence in the sediments, i.e., the rocks, that are now photographed by rovers. Furthermore, microbes and microalgae are the first step on the evolutionary scale of life on Earth and stromatolites are the oldest evidence for them, stretching back at least 3.5 billion years. Hence, these are the structures that may be present if life ever existed on Mars comparable to Earth [19].

The presence of extraterrestrial microorganisms and, in particular, of cyanobacteria, well known as the main builders of terrestrial stromatolites, has been suggested by many authors beginning from the famous discovery of Martian meteorite ALH84001 [27]. This biological approach was further confirmed by some intriguing images of the Martian surface sent by rover Opportunity on 2010, showing a set of rocks partially covered by a dark shiny patina, close to the terrestrial "Desert Varnish" probably formed by the same bacteria that built stromatolites on the Earth [28]. The latest finding is the discovery in CL1 carbonaceous Martian chondrite of some microfossils very close to terrestrial cyanobacteria [29]. All these hypotheses were strengthened by some studies about extremophiles as desert cyanobacteria of the genus Chroccocidiopsis living in extremely hot and cold deserts that sheds a new light on the possible history of Martian microbial life [30, 31]. In particular, it has been pointed out the important role of Cyanobacteria in the formation of organo-sedimentary rocks as one of the most successful and widespread forms of life on Earth [32] because of their great morphological variability and their bio-stabilization capacity on sun-light exposed sedimentary surfaces, but also in environments characterized by extremely low energy light. In this way, the recent discovery of a new photopigment, found within terrestrial stromatolites and named Chlorophyll f, that can absorb light of even lower photon energy until 720 nm [33], suggests that cyanobacteria could alive also in extreme environments as on Mars. Generally, these microorganisms occupy a very broad range of environments including waters of widely different chemistry and composition so that their involvement in sedimentation processes is equally varied [34]. Cyanobacteria, including more than 2000 species [35], as composite microbial associations, dominate microbial mats and are ubiquitous, leaving successful records in sediments and sedimentary rocks.

Morphological study of images reveals evidence of widespread occurrence of micro, meso, and macro structures recalling for some authors early terrestrial forms of life; such as the "blueberries", concretions possibly induced by chemolithoautotrophic bacteria [10, 36, 37]. These strange and complex structures, for which abiological explanation it's hard to find, have strong morphological parallels with terrestrial microbialites/stromatolites [14–16, 18], a conclusion that seems to be supported by morphometric approaches [12, 13]. Other possible biogenic structures have been observed on Mars, recalling those of terrestrial silica deposits in hydrothermal environments [38, 39] or typical structures, known as Microbially Induced Sedimentary Structures (MISS) and generated by microbial mats of intertidal environments [14]. Despite the many observations, mutually supporting a possible microbialite hypothesis, they do not prove the presence of fossil life on Mars, because biologic explanations for their terrestrial counterparts and for their contained microbial structures are often deeply controversial. In fact, such organo-sedimentary structures are sediments, and despite having unusual features at meso and macroscales, somewhere contain controversial microbial remnants of micrometric dimensions; while complex and larger structures, as are evolved and more evident microfossils (generally larger than 0.1 mm) or macrofossils (centimetric) are generally more obvious and indisputable. Finally, in the Martian atmosphere it has been detected traces of methane and formaldehyde, changing seasonally and supporting evidences on the potential habitability of Mars [40].

## 3. Intriguing sedimentary microstructures, chemicals and complex organics found by Curiosity rover at Gale Crater

At Pahrump Hill (Mojave and Mojave 2 targets, at Sols 809 and 880) a large number of light-toned lozenge-shaped pervasive microstructures having dimensions and shape comparable to rice grains (nicknamed here as "rice grain") were found and regarded by NASA as pseudomorphic sulfate crystals resulting from occasional lake evaporation [7]. In this location, Curiosity detected mineral assemblage suggest, paradoxically, both oxidizing (hematite) and reducing (magnetite) environments, as well as acidic (diagenetic and/or authigenic jarosite) and neutral (apatite) conditions; findings difficult to explain as having an abiological origin and more compatible with a biological origin.

In this work we hypothesize that among the structures imaged by Curiosity at Mojave targets some are suggestive of bacterially mediated reactions or are the remnants of fossilized life forms, and which also agrees with the organic molecules detected just on such target (**Table 1**).

Hence, the objective of this study was to investigate the microstructures observed at Sols 809 and 880 (Mojave and Mojave 2 targets; the same site on successive paths) in order to determine their possible biogenicity. These structures were described by NASA as lozenge shaped, pseudomorphic minerals (https://www.nasa.gov./jpl/msl/pia19077).

First, we studied the morphology of microstructures with reference to possible parallels with primitive terrestrial forms of life, such as stromatolites/microbialites, microfossils and/or algae, taking in account previous similar possible finds observed by Martian rovers. Parallels emerged by a systematic image analysis, considering terrestrial similar forms and sedimentary processes, both syn-genetic and post-genetic, in a given environment.

Second, we morphometrically analyzed the lozenge-shaped structures, here called "rice grains" (**Figure 2**), observed at the same targets, in order to investigate if their microstructures are compatible to sulphate crystals (Gypsum, Jarosite) as well to other minerals contained in these outcroppings or to primitive terrestrial forms of life.

## 4. Materials and methods

### 4.1 Morphology

The study was based on Mahli (Mars Hand Lens Imager), ChemCam and Mastacam images uploaded on NASA web site (https://mars.nasa.gov/msl/multimedia/raw-images), and especially those from the Sols range 750–1400. A selection of images (**Table 2**) was made which were considered particularly interesting and have been morphologically analyzed in detail by amplification and color/brightness/contrast modifications, in order to evaluate possible comparison to analogous terrestrial form of life. In particular, Mahli employs a macro lens color camera able to focus on targets at working distances of 2.1 cm to infinity, with a maximum resolution of about 14 μm/pixel. We have used a scale for Mahli images using the relation between motor count and the pixel dimension values provided by NASA and whose pixels are in the range from 20 μm to 30 μm. Consequently, the dimensions of microstructures analyzed in this article are, at least, about 10 time higher than image resolution.

| |
|---|
| Methanethiol ($CH_4S$) |
| Dimethyl sulfide ($C_2H_6S$) |
| Thiophene ($C_4H_4S$) |
| Methylthiophenes ($C_5H_6S$) |
| Benzothiophene ($C_8H_6S$) |
| Carbon-chain molecules with 1 to 5 carbons |
| Benzene ($C_6H_6$) |
| Chlorobenzene ($C_6H_5Cl$) |
| Alkylbenzenes/benzoate ion ($C_8H_9/C_7H_5O^-$) |
| Toluene/tropylium ion ($C_7H_8/C_7H_7^+$) |
| Naphthalene ($C_{10}H_8$) |

**Table 1.**
*Carbon-containing compounds revealed by Sample Analysis at Mars (SAM) aboard of Curiosity rover: thiophenic, aromatic and aliphatic molecules in Mojave target.*

## 4.2 Morphometry

### *4.2.1 Euclidean morphometry: comparison of "rice grains" with various crystals and algae*

The purpose of this section was to investigate quantitatively the features referred by NASA as light toned lozenge-shaped microstructures (**Figure 2**). This investigation determined, using a series of metrics, including degree of dispersion, variability in lengths and widths, length/width ratio, fit of lengths to a log-normal distribution, fit of orientations to a rectangular distribution, and morphological analysis, the similarity between the "rice grains" and various abiotic mineral deposits and terrestrial life forms. In particular: images of Gypsum and Jarosite crystals, an image of Feldspar phenocrysts, and a population of terrestrial *Euglena viridis* (O. F. Müller) Ehrenberg, 1830, (**Table 2**) were observed. These images were analyzed using 'Image J' software developed by the National Institute of Health (NIH), Bethesda, USA [42–44].

Each image was magnified to clearly reveal the objects of interest. Images were manipulated using brightness, contrast, sharpening and edge detection to optimize the appearance of the objects and to establish their boundaries. A grid of squares was then superimposed over each image to establish a number of sample fields. Each of the "rice grains", with at least 50% of its area within a sample field, was measured. Various relative measures based on degree of variation, ratios, fit to various distributions and proportions exhibiting a specific morphology, were used because scale measures were not always available for all images or those that were quoted were unreliable.

The following data were obtained from each sample field containing "rice grains": (1) total number of profiles, (2) the maximum length of each profile in relative units, (3) the width of each profile taken at the midpoint, (4) the orientation of each profile, measured as the angle between the horizontal and a line drawn along the maximum length of the profile, (5) the proportion of the profiles which had a fusiform shape and (6) the proportion which exhibited a degree of flexibility or curvature relative to a straight line drawn connecting the two ends of the profile. A number of metrics based on the profiles were analyzed and compared: (1) spatial

**Figure 2.**
*A set of amplified image samplings (rover image MAHLI, taken at Sol 809) showing chaotically arranged whitish forms, many of which are bezel or fusiform-shaped bodies. Such structures were interpreted by NASA as lozenge-shaped crystals (report PIA 19077; central image). Samplings were obtained by using contrast adjustments of 40% (above) and 30% (below), and the more appropriate luminosity, case by case.*

pattern, i.e., whether the profiles were distributed at random, uniformly, or were clustered, (2) degree of variation in length and width, measured as the coefficient of variation (CV) (3) size frequency distribution of lengths, (4) size frequency

| Figure | Section | Type | Sol | Source |
|---|---|---|---|---|
| 4 | A,B | Terrestrial: Stromatolite | — | Web site |
| | C | Mars: Concretions | 767 | Curiosity |
| | D | Mars: Concretions | 871 | Curiosity |
| | E,F | Mars: Concretions | 758 | Curiosity |
| 5 | A′B′ | Terrestrial: Microbiolites | — | Web site |
| | A | Mars: Laminate structures | 890 | Curiosity |
| | B | Mars: Laminate structures | 810 | Curiosity |
| 6 | A,D,F | Mars: Filaments | 871 | Curiosity |
| | B | Mars: Filaments | 758 | Curiosity |
| | C,E | Mars: Filaments | 899 | Curiosity |
| | G | Mars: Filaments | 598 + 715 | Opportunity |
| | H,I | Mars: Filaments | 780 | Curiosity |
| 2 | All | Mars: 'rice grains' | 809 | Curiosity |
| 8 | All | Mars: 'rice grains' | 880 | Curiosity |
| 9 | Up, Lw L | Mars: Various morphologies | 880 | Curiosity |
| | Up inset | Terrestrial: *Euglena* | — | G. Bianciardi |
| | Lw R | Dasycladales-Epimastopora(f) | — | Web site |
| 10 | A,B,D,E | Mars: 'conical bodies' (f) | 1103 | Opportunity |
| | C | Terrestrial: Dasyclad (f) | — | Web site |
| 11 | A,B,C,L,G | Terrestrial: Dasyclad fragments(f) | — | Web site |
| | D,E,F,H,I,M | Mars: Shell fragments (f) | 1273 | Curiosity |
| | O,Q | Mars 'rice grains' | 880 | Curiosity |
| 12 | 1 | Terrestrial: Stromatolite filaments (f) | | [63] |
| | 2 | Terrestrial: *Gymnocodium* (f) | — | [41] |
| | 3 | Terrestrial: *Euglena viridis* | — | G. Bianciardi |
| | 4 | Terrestrial: Gypsum | — | V. Rizzo |
| | 5 | Terrestrial: Jarosite | — | (jarosite4138d, www.dakotamatrix.com) |
| | 6 | Terrestrial: Feldspar | — | (1200px-granite-phenocrysts, it.wikipedia.org) |
| 14 | A,B,C | Mars: Filaments | 880 | Curiosity |
| | D | Mars: Filaments | — | Opportunity |
| | L | Terrestrial: Algae and Cianobacters | — | A. Munneke |
| 15 | — | Terrestrial: tubular septate bodies | — | Hong Hua |

| Figure | Section | Type | Sol | Source |
|---|---|---|---|---|
| 16 | 1–15 | Terrestrial: Euglena | — | Hong Hua |
| | A,B | Mars: 'rice grains' | 880 | |

**Table 2.**
*List of terrestrial and Martian images used in the analysis of putative microalgae on Mars. (Abbreviations: Up = Upper, Lw = Lower, R = Right, L = Left, F = Fossil).*

distribution of orientations, (5) overall shape of the profiles, i.e., whether fusiform or not fusiform and whether a degree of curvature was present. A Poisson distribution was fitted to the objects from all images. If the objects were distributed at random, then, the probability (P) that the fields contain various numbers of profiles is given by the Poisson distribution [45]. The variance (V) of a Poisson distribution is equal to its mean (M) and hence, the V/M ratio is unity for a random distribution. The V/M ratio can, therefore, be used as an index of "dispersion", uniform distributions having a V/M ratio less than unity and clustered distributions greater than unity. In addition, profile diameters and orientations were used to construct size class-frequency distributions. Two statistical models were fitted to these distributions: (1) a log-normal distribution [46], often used to describe the size distributions of plant populations [47, 48] and (2) a rectangular distribution, to test whether the objects exhibited any preference with regard to orientation. Goodness-of-fit to the various models was tested using the Kolmogorov–Smirnov (KS) test. To study the similarities among the ten images, the data were analyzed using Principal Components Analysis (PCA). The analyses were carried out using the images as variables and the various metrics as defining features. The result of PCA is a scatter plot of the images in relation to the extracted Principal Components (PC) in which the distance between the images reflects their relative similarity or dissimilarity based on the defining metrics. To correlate the location of an image on a PC axis with the specific metrics, correlations (Pearson's 'r') were calculated between the values of each metric from each image and the factor loadings of the case relative to the PC1 and PC2. For example, a significant correlation between a specific metric and PC1 would identify that feature as particularly important in determining the separation of cases along PC1.

### *4.2.2 Fractal morphometry: comparison of "rice grains" with various crystals and algae*

"Rice grains" images were enlarged 10 x by Paint Shop software in order to reach the final size of the sample.

Images were then loaded on paint.net software (https://www.getpaint.net/download.html) in order to extract the specimen of interest. The extracted image was loaded on Digital Image Magnifier (Nikolao Strikos, https://sourceforge.net/projects/digitalimagemag/) in order to apply a Canny edge filter to the image, fixed Sigma, High and Low thresholds were used.

A negative of the skeletonized image was then obtained. The skeletonized image was loaded on Benoit Fractal Analysis software (https://www.trusoft-international.com/) in order to evaluate the Fractal Dimension of the image: Fractal dimension, D0, a measure of the space-filling properties of a structure, was calculated on the skeletonized images by the box-counting method. Briefly, each image was covered by nets of square boxes (from 5 to 100 pixels) and the amount of boxes containing any part of the outline was counted. A log–log graph was plotted on the side length of the square against the number of outline-containing squares, the slope of the

*DOI: http://dx.doi.org/10.5772/intechopen.95531*

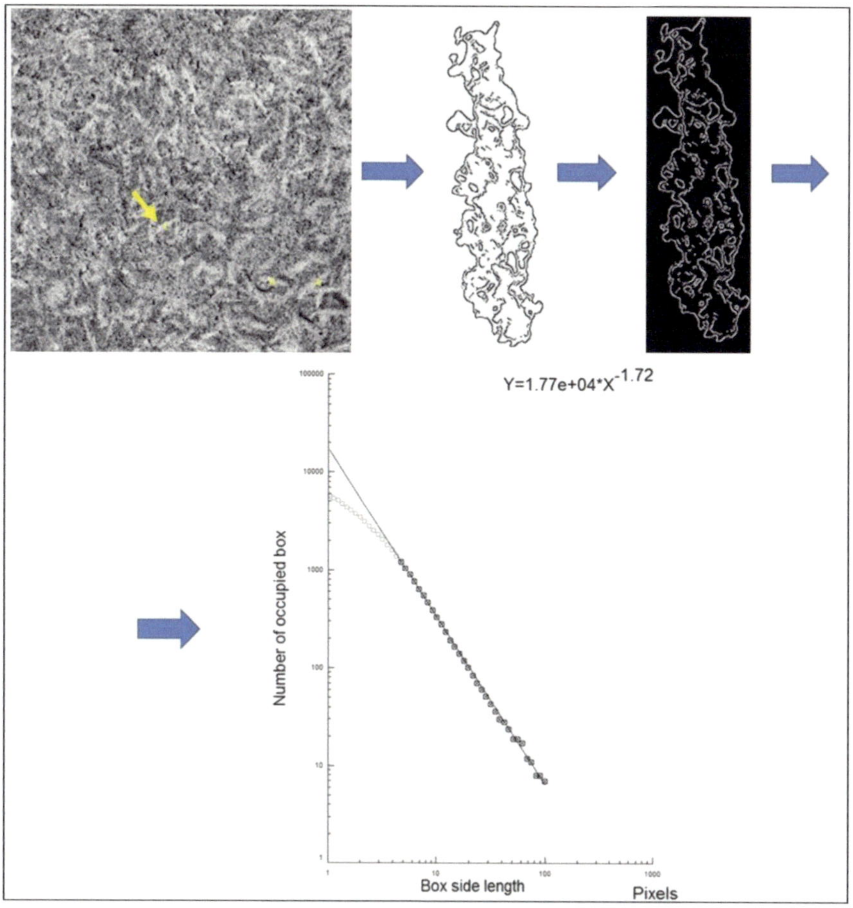


**Figure 3.**
*Top: From a Martian sample (rover image MAHLI, taken at Sol 809, "rice grains"), a single "rice grain" is enlarged, a canny filter is applied and the negative obtained. Bottom: the log log plot is a straight line ($p < 0.01$): the "rice grain" owns a fractal structure or self-similarity, its exponent is the fractal dimension.*

linear segment of the graph representing the local fractal dimension of the image (**Figure 3**). The linearity of the log log plot was assessed by the Pearson's correlation test, in order to demonstrate that the "rice grains" owns a fractal structure and fractal dimension may be performed. Variance analysis was used in order to compare fractal dimension of the Martian "rice grains" vs. the mineral (abiologic) gypsum or versus *Euglena viridis*, as a model of a complex (eukaryotic) microorganism.

## 5. Results

### 5.1 Environmental and morphological evaluations.

#### *5.1.1 Sedimentological context and possible biogenic structures*

Curiosity landed on a flat plain (the Bradbury landing site) to the North of Mount Sharp in August 2012. Subsequently, the rover travelled to reach the extensive strata of a lacustrine sedimentary sequence at the base of Mount Sharp (around Sol 750), detecting along this track a heterogeneous assemblage of sedimentary rocks, representing a fluvial-deltaic-lacustrine environment (the Yellowknife Bay formation). The basal Sheepbed and Gillepsie members of the Yellowknife formation are characterized by silts and mudstones showing mutual unconformity; such sediments were deposited in an intertidal region of a shallow lacustrine environment, representing the latest stage of transport and deposition of fan sediments inside Gale Crater lake [1].

In the first part of the Curiosity survey, interesting chemical and structural data emerged. Hence, with reference to the Sheepbed formation, at John Klein and Cumberland sites, SAM detected organic chemicals referable to molecules of chlorobenzene as well as the occurrence of all the chemical components of life [49]. Very interesting structures and morphologies were also observed, in addition to the many already cited, possible microbialites on the Sheepbed mudstone and structures recalling terrestrial microscopic induced by sedimentary structures, known as MISS, locally present as "erosional remnants," "pocket," "mat chips, "roll ups," "desiccation cracks," and "gas domes" [14]. Other possible microbially induced structures were also observed at this site, e,g., burst bubbles, filaments, mini-atolls, oncoids, domes and many other atypical sedimentary structures known as microbialitic [17, 39, 50], and some of which resemble stromatolites of the "Conophyton" type [18]. However, the morphology together with the chemical data mutually support the presence of ancient life, even if analyzed individually, they suggest alternative abiotic interpretations. For this reason, they will be individually considered, according to the criteria established by Astrobiology Field Laboratory (AFL) at http://mepag.jpl.nasa.gov/reports/index.html, as "possible biosignatures".

All previous described putative organosedimentary macrostructures were not observed when Curiosity moved away from the intertidal zone and entered (at Pahrump Hills location) a clay sedimentary sequence, the Murray formation at the base of the Mount Sharp. This was an area well investigated by the rover between Sols 750 and 930, and where deposits of mudstone and siltstone outcrops reveal a very thin, sub-millimetric lamination, proving the occurrence of a hydrodynamically stable environment. Particularly worthy of attention are the very thin laminated outcroppings which contain a few widely distributed aggregated harder structures [51], some up to several cm in size (**Figures 4** and **5**). Such dendritic, nodular and laminate concretions, comparable to the host rocks, show notable Mg-enhancement and a strong depletion of other major elements [52].

Chemical results indicate that the Gale Crater sediments were strongly influenced by early, subaqueous diagenetic reactions that produced, and sometimes filled a variety of pore types [53]. Since some sulfur was also detected in a dendritic feature, Mg occurrence was likely to be associated with a $MgSO_4$ phase: a mineral present as precipitated cement within sediment pores [2] and an indicator of a local very acid environment, confined to the hardest structures and, intriguingly, in a target (Mojave) where organic molecules, such as kerogen and one of its polyaromatic fragments (thiophene), were found [54]. Considering the lack of contact with host sediment, these harder structures appear not to be transported/deposited elements but diagenetic structures, formed after sediment deposition. In fact, the two step heating experiments made by Martin to date K-Ar ages of primary (at 930°C) and secondary (at 500°C) mineralogical components at Mojave 2 target gave respectively 4.07 +/− 0.03 Ga (associated to detrital plagioclase of lake sediments) and 2.12 +/− 0.36 Ga (associated to sulphates and other secondary components of the diagenetic structures including the "rice grain"). Such data also suggest a post 3 Ga aqueous phase occurred in Gale Crater, at a time after surface fluvial activity on Mars was thought to have largely ceased [7].

Hence, at the landing site of Gale Crater, conditions for the evolution of Martian life beginning with chemotrophic and anaerobic bacteria and which survive in fossilized traces has been hypothesized [11]. Moreover, chemical and textural data

**Figure 4.**
*Concretions found on the Murray Formation, Mars (C-F) in comparison to some Permian lacustrine terrestrial stromatolites (frames A,B; from: https://www.paleodirect.com/strx005-permian-lacustrine-stromatolite-colony-fossil). Both Martian samples and terrestrial stromatolites counterparts show harder structures consisting of lumps/nodules, that developed in close proximity (B, D) or became randomly aligned forming a branched (F) or overlapping structure (C), starting from single point (F and C) or from scattered "germination points" (D-F). One can also observe similarities on their surface pattern (B-E yellow circles) and distribution.*

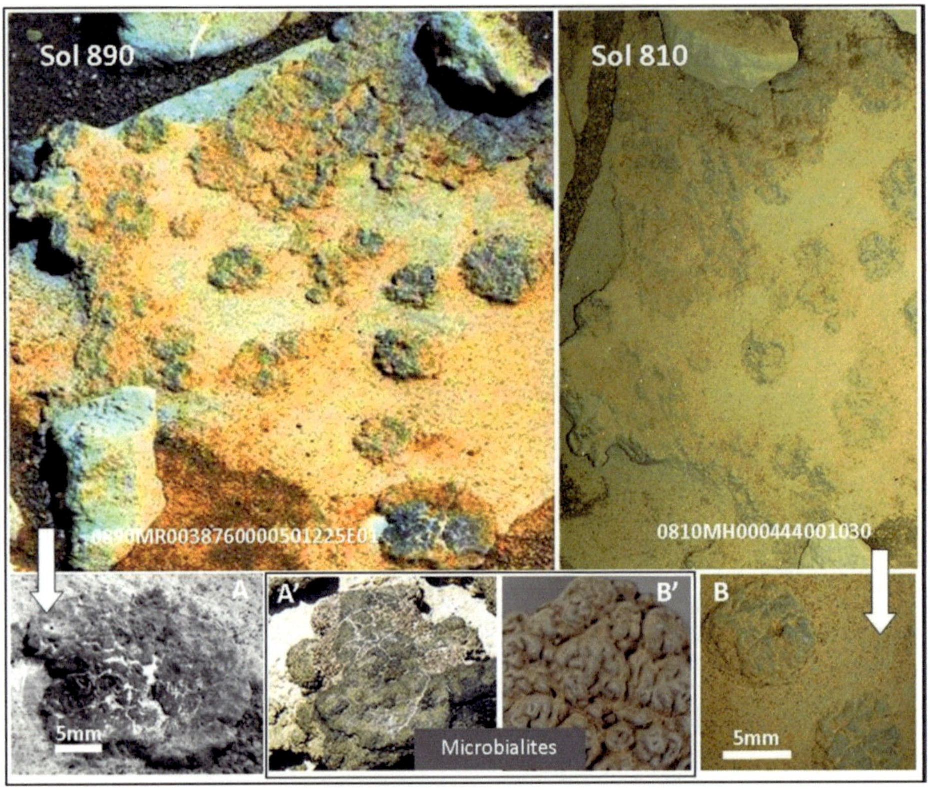


**Figure 5.**
*The two upper images (taken at the same outcropping, at sols 810 and 890), show irregularly hardened lamination. This case (irregularly developed in layers) is conceptually similar to those shown on* ***Figure*** *2 and shows local transition to nodules (A, B). For this reason these hardened structures may suggest a syngenetic or early diagenetic origin, due to different spatial activity of colonies, rather than the effect of erosion. Note the morphological similarity to terrestrial microbialites (frames A-A' to an encrusting thrombolites; frames B-B' to a cerebrotic surface of stromatolites).*

testify to the presence of almost neutral lake water and to considerable diagenetic variations in pH [52]. The latter may suggest in very locally confined areas, that the acid micro-environment could also be attributed to microbial activity. Hence, the presence of elongated and curved filaments inside these structures and on their surface both in terrestrial stromatolites and Martian samples is particularly noteworthy (**Figures 6** and 7).

### *5.1.2 Nodular/dendritic harder structures*

The harder structures may result from small grain aggregates, randomly dispersed, preferably along diverging alignments from scattered points. As a consequence, they assume several unusual shapes, forming lumps and nodules, and occasionally branched and/or overlapping bodies (called "dendritic"; [55]). The same structures also occur in the laminated sequences typically forming irregular crusts and emerging nodules, marked by an irregular scabrous surface, typical of stromatolites (known as "cerebrotic") and thrombolytic crusts. In these structures and on their surface, both in terrestrial and Martian samples, elongated and curved filaments are sometimes noted (**Figures 6** and 7); most noteworthy, the elongated structures of Martian samples, occasionally exhibit regular septate forms (**Figures 6–9**).

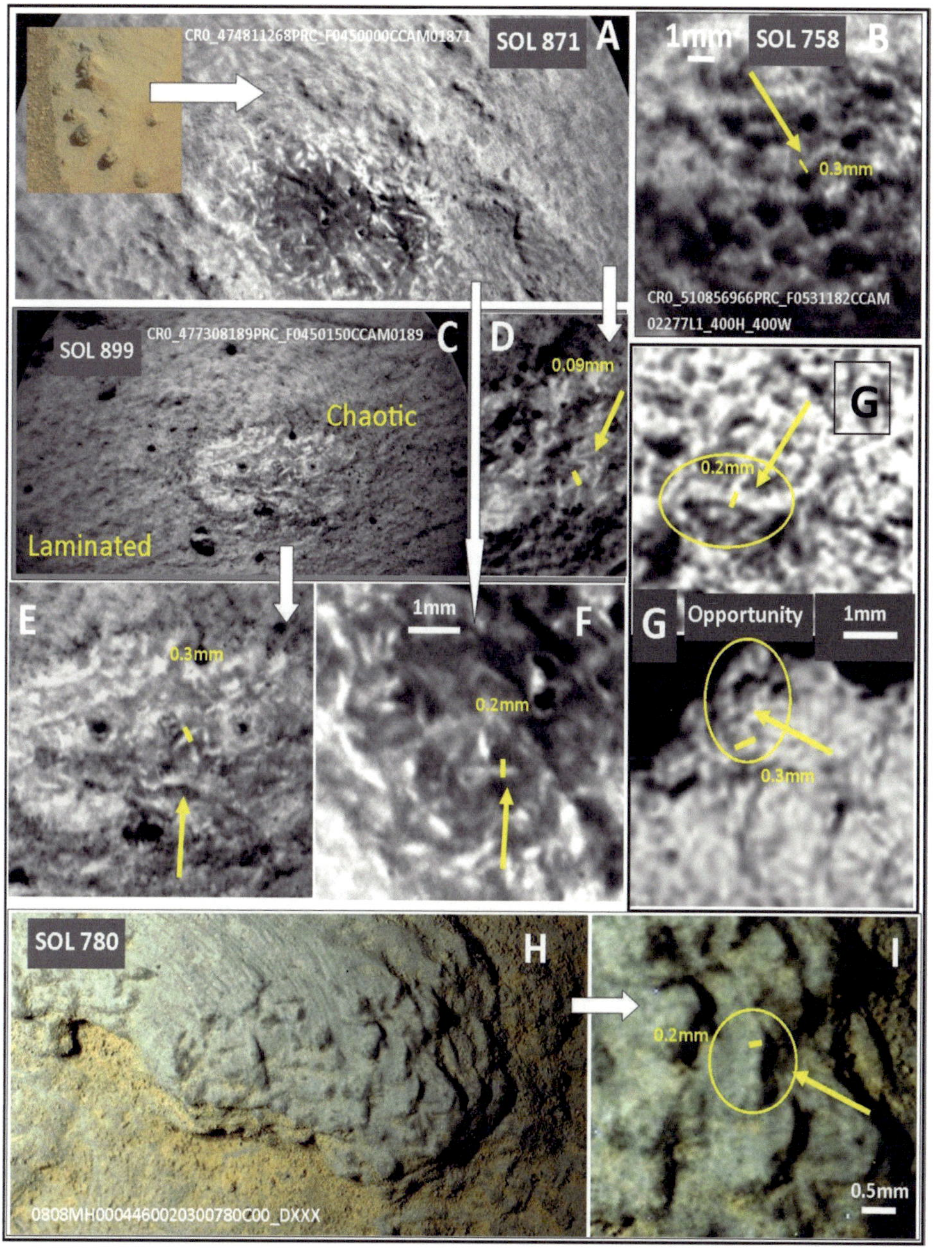


**Figure 6.**
*These images show a series of filamentous/elongated microstructures taken by MAHLI (H,I) and ChemCham (A-B) in the hard concretion of the Murray Formation, in comparison to other Martian filamentous structures taken by Opportunity rover (G). The selected elongated microstructures (arrows and yellow circles are in relief in all frames); they show cross sections of 0.09–0.3 mm and regularly septate interspaces (more evident in G, H and I), forming elongated sinuous (D-I) and intertwined structures (E-I). On note the lateral discordance, from a laminated setting to a more chaotic structure (C), common for terrestrial microbialites.*

Such dendritic, nodular and laminate concretions have been investigated in detail by the NASA scientific team [52]. They found, comparable to the host rocks, notable Mg-enhancement and a strong depletion of other major elements; nickel also being reported by ChemCam. Since some sulfur was also detected in a dendritic feature, they interpreted the Mg occurrence as of the presence of $MgSO_4$ phase; a

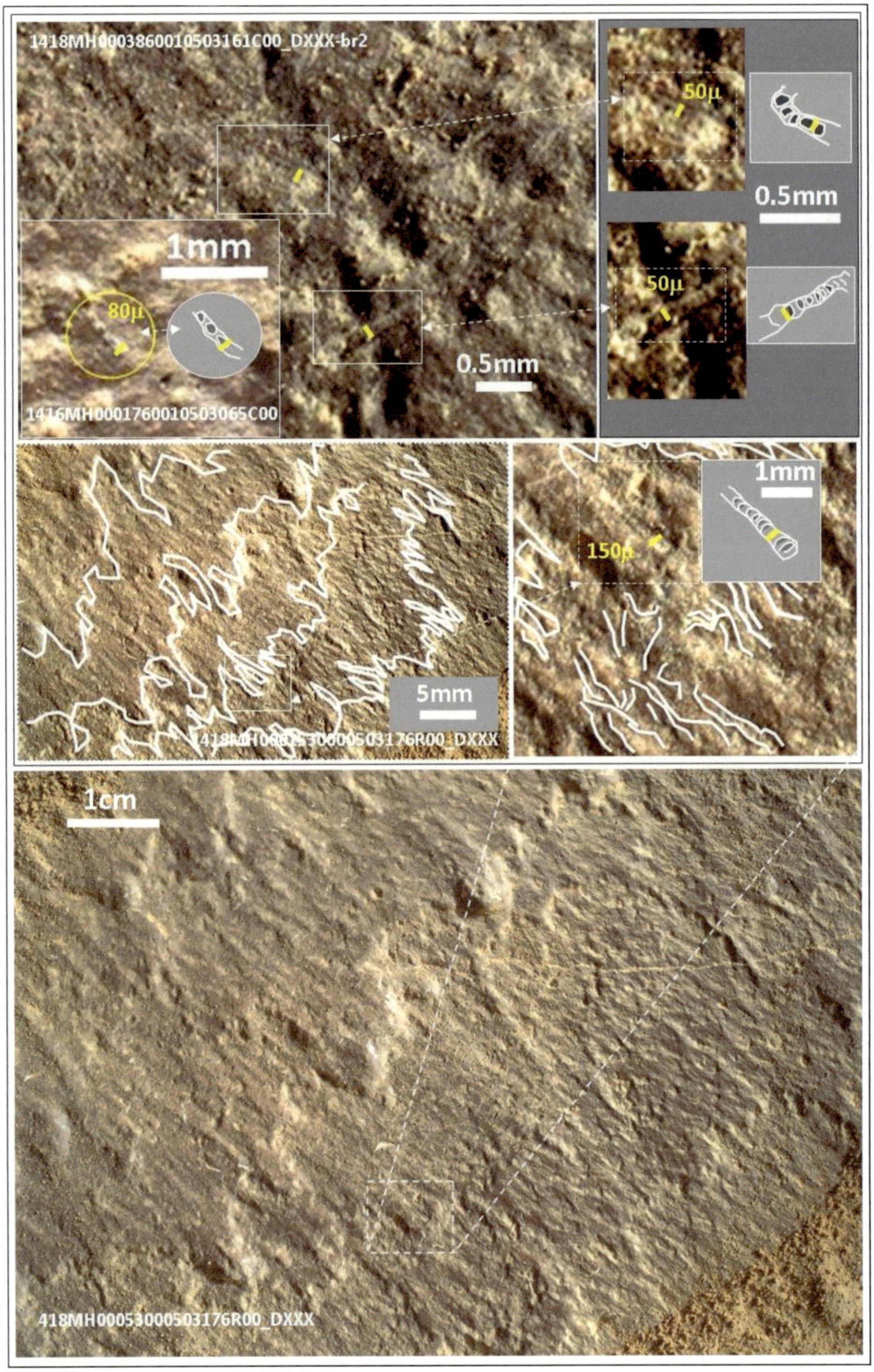


**Figure 7.**
*Harder septate filaments forming thin laminae (Murray Formation; Sols 1416–1418). Images show a series of filaments (as shown in sub-frames) having transverse dimension ranging from 50 to 150 microns. Such filaments stand out from the rock and show a sequence of aligned harder segments, having septate bodies. Their shapes are often sinuous, several millimeters long; in the above images they are in relief, orientated NW-SE, thus determining, due to their hardness, serrated contact between the laminae (white lines).*

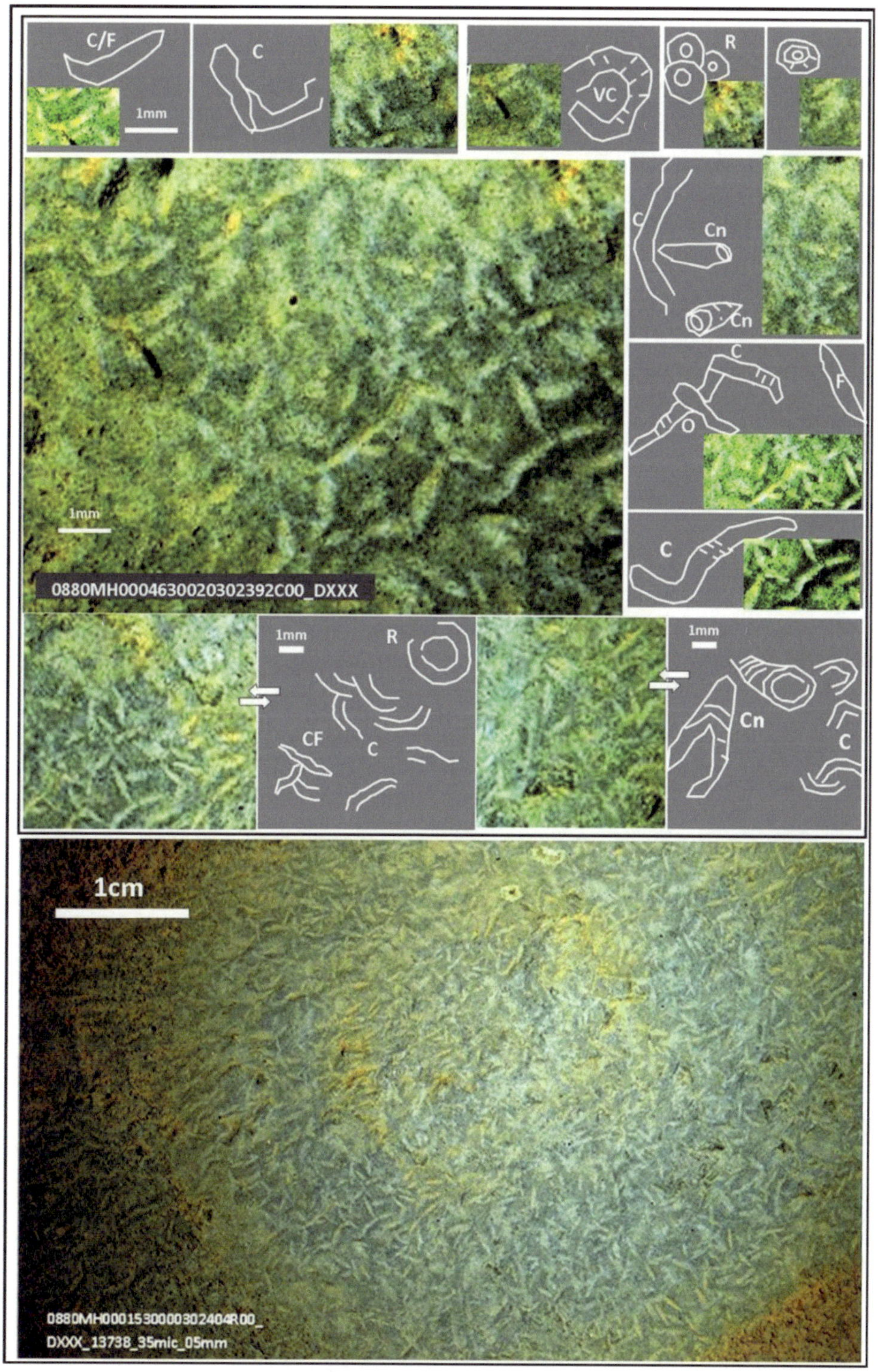


**Figure 8.**
*Morphological interpretation of some of the "rice grains" on an amplified MAHLI image (Sol 880). One can see repetitive curved (C), overlapping (O), fusiform (F), conical (Cn), very curved (VC), ring-shaped (R), curved/fusiform (C/F) shapes and therefore showing significant differences from the expected regular appearance of crystals. Note the presence of shapes (R and Cn types) which we have interpreted as possible transverse sections of conical bodies.*

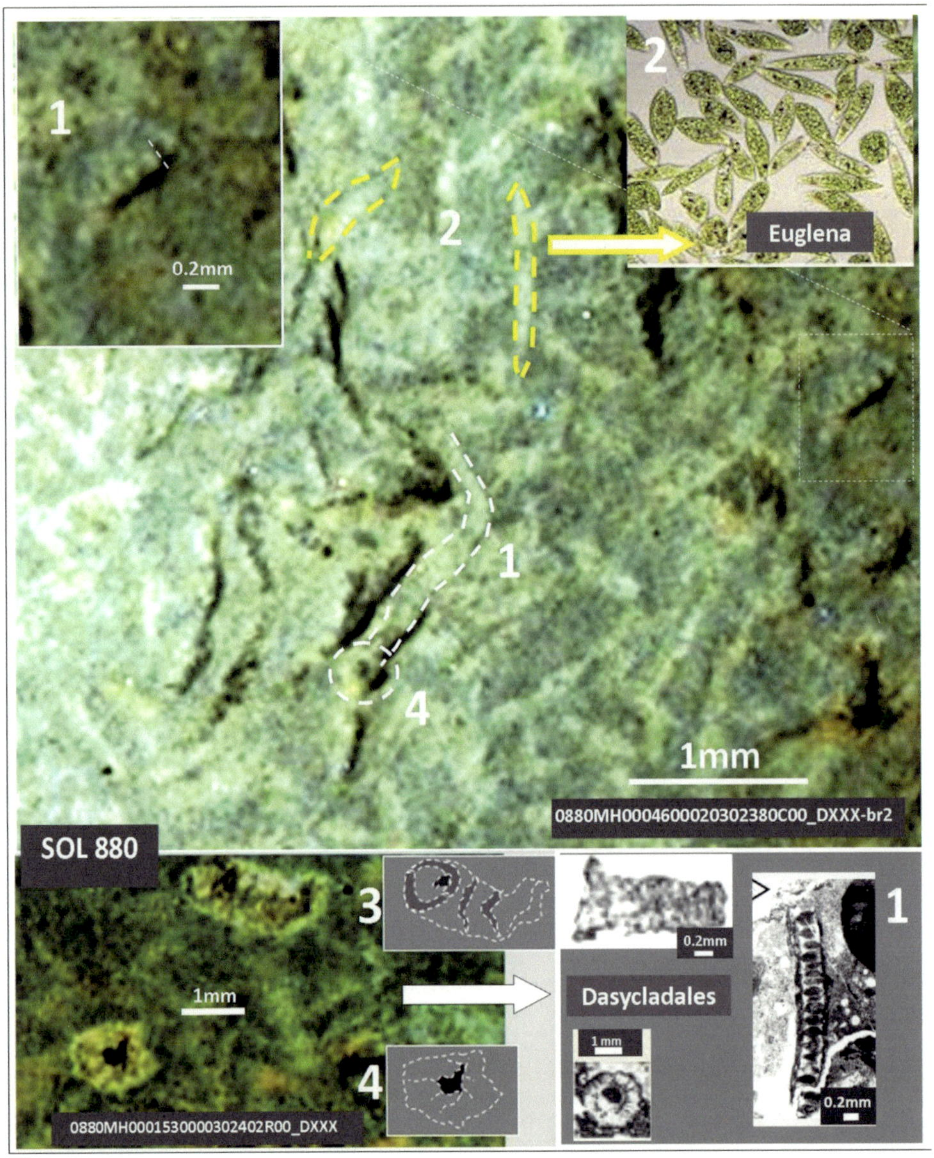


**Figure 9.**
*Highlighting the most common forms observed in MAHLI images at Sols 809 and 880. The amplified image in the center shows the following types: elongated/curved body (1), occasionally septate (frame left on the top; see dashed line), lenticular/curved (2), holed conical (3) and possible transverse sections of previous type bodies (4). Examining possible terrestrial analogues, such microstructures are similar to terrestrial Dasycladales, Euglenoid, or giant filamentous cyanobacteria, rather than crystals. Please note that, the shapes are unambiguously identified by the large number of colored pixels contained.*

mineral present as precipitated cement within sediment pores [2], and hence an indicator of a local very acid environment and confined to the hardest structures.

Information regarding the dimensions of such structures, their shape and their layering, in relation to the sedimentary environment and their possible origin, are particularly important. Structures appear to be embedded within sediments of a quite aqueous environment and lack net contacts. Despite their random form and distribution, they show common features and appear as irregular aggregates, composed of globular and/or linear structures, affecting groups of laminae.

The dimension of these aggregates, their unusual structure and distribution and the lacking of net contact with the host sediments, that'suggests that they are not transported/deposited elements, but that their formation was inside the sediment, after deposition, due to local cementation and/or grain rearrangement; conditions likely to be present during diagenesis and still existing in lake waters. In fact, chemical results indicate that Gale Crater sediments were strongly influenced by early, subaqueous diagenetic reactions that produced, and sometimes filled, a variety of pore types [49, 53]. Chemical and textural data indicate almost neutral lake water, while considerable local variations in pH [52]; so, they may suggest micro-environments and related microstructures generated by microbial activity and/or to the presence of local organic material. In addition, at the Mojave target site, traces of thiophene were found by the NASA scientific team, one of the main elements of kerogen; an organic compound that may be related to bacterial metabolism associated with terrestrial microbialites and, commonly, used as one of the main criteria to assess the biogenicity of putative Archean stromatolites [54, 56]. Particularly worthy of attention, both internally and externally, terrestrial stromatolites and Martian samples, both show elongated and curved filaments (**Figures 6** and 7).

### *5.1.3 Filamentous microstructures*

Occurrence of filamentous structures, detectable by a different color and tone variation, normally appear to be formed by septate bodies having transversal dimensions of 0.05–0.3 mm (**Figures 6** and 7). These structures are more evident on amplified/slightly blurred images of Martian sediments and on clean/abraded surfaces. Occasionally, they are more resistant and in relief, conditioning the shape of the laminar surface and their mutual contacts (**Figure** 7). They were observed, both as single features as well as intertwined structures and cover the rock surface resulting in a "woven" texture. Similar settings (**Figure 6**, frame G) have been investigated in previous work and interpreted, by visual and numerical approaches on a consistent number of terrestrial sampling analogues, as microbial/stromatolitic structures [12, 13, 17].

### *5.1.4 Lenticular and conical/tubular structures*

MAHLI images taken at Sol 869 show that the lenticular lozenge-shaped "rice grains" observed on brushed surfaces at Sols 809 and 880 (Mojave target), not only occur "on the surface" as harder and whitish structure, but massively affect the entire outcrop, covering about 50% of the lithological mass. Previously, these structures have been interpreted as mineral deposits, e.g. of Gypsum or Jarosite (NASA reports), but subsequently, due to the lack of sufficient amount of crystal, they were interpreted as pseudo-morphic crystals originating from amorphous substances.

In particular, the mineralogical composition of Mojave 2 (Sol 880) shows, in respect to the previous investigated rocks, a variation in mineral composition, exhibiting significant amounts of amorphous material (54%) and minor amount of Plagioclase (24%), Magnetite (4%), Hematite (4%), Jarosite (4%), Phyllosilicates (5%) and Fluorapatite (5%). Such data paradoxically suggest the coexistence of both oxidising (Hematite) and reducing (Magnetite) environments, as well as acidic (Jarosite) and neutral (Fluorapatite) components. Chemical data of this sample suggests the following composition: $SiO_2$ (49%), FeO (16%), MgO (4%) CaO (4%) and $Al_2O_3$ (11%), together with other minor components, including Magnetite and Phosphorous [2].

Some of these minor components and minerals, such as Apatite, Magnetite, Ni, Zn and Br (from Curiosity APXS results; in [7]), found at Mojave target, on Earth are generally associated with microbial activity and stromatolites [50, 57] and suggested by AFL report as possible biosignature [58].

Given the lack of mineralogical (CheMin) or chemical (APXS) evidence for calcium sulphates in the Mojave 2 sample (Sol 880), it was assumed that these lenticular bodies represent, on the basis of their morphology, i.e., lenticular gypsum crystals laths, light color compared with the host rock, and penetration vertically into the bedrock, crystals laths that were formed syndepositionally with the Murray mudstone and were later re-dissolved by post depositional fluid flow, forming pseudo-morphic microcrystalline or amorphous substance of unknown composition [2, 7].

It should be noted that the occurrence of microbially precipitated fluorapatite is reported in Jurassic phosphate stromatolites by Sànchez [57] and it is also known that biomineralization processes could give Biologically Induced Mineralization (BIM) and Biologically Controlled Mineralization (BCM), where magnetite is one of major components [59]. Moreover, such structures were found in association with the previously described problematic diagenetic features [55, 60, 61], and hence, a number of controversial features are suggestive of possible biogenic shapes.

Morphological analysis of MAHLI images at sols 809 and 889 (at Mojave1 and Mojave2, respectively) reveal that the "lozenge-shaped sulphates" [51] show chaotic, mainly fusiform/filiform, septate, curved shape; some of which are in relief and resemble terrestrial microalgae (**Figures 8** and **9**).

In particular, considering their shape and dimension, we investigated the structural similarities with Dasycladales algae, giant filamentous Cyanobacteria or Euglenoids. This biological interpretation could be supported by the occurrence of two adjacent 'bright' bodies, present in the same image (Sol 880; **Figure 9**, features 3 and 4). Occurrence in the same target of spherical cross sections (having sharp inner surface and irregular outer edge), could be related to other cones, although less evident and in small amounts (**Figure 8**, Cn and R features). Another conical body, photographed by Opportunity at Meridiani Planum, shows a differentiated skeletal structure in transverse section (a conical thallus?), and possible regular radial laterals attached (of aspondyl type? **Figure 10**). In addition, images at Sol 1273 show transverse and oblique sections of a tube-like bodies associated with regularly jagged discontinuities of their shells (**Figure 11**; features 1–3). Hence, the variety of shapes, present which include septate filamentous structures, is of great biological interest and do not suggest crystal-type structures; and are worthy of morphometric investigation (**Table 2**, **Figure 12**), the results of which are reported in the following section.

## 5.2 Morphometric results

### *5.2.1 Euclidean morphometry*

The metric data for each image is summarized in the upper section of **Figure 13**. There was considerable variation in all metrics among images with the exception of the fit of lengths to a log-normal distribution, the KS tests suggesting that this distribution fitted the objects of interest in all images. V/M ratio varied from a maximum value of 1.08 (Jar-1) indicating a random distribution of profiles, to a minimum of 0.32 (Euglena) suggesting the majority of profiles exhibit a degree of uniformity in their distribution. Variation in length and width also varied among images being least in the Martian sample and Terrestrial Euglena and greatest in the alga Gymnocodium. Length/width ratios were greatest for Gymnocodium and least for the three gypsum

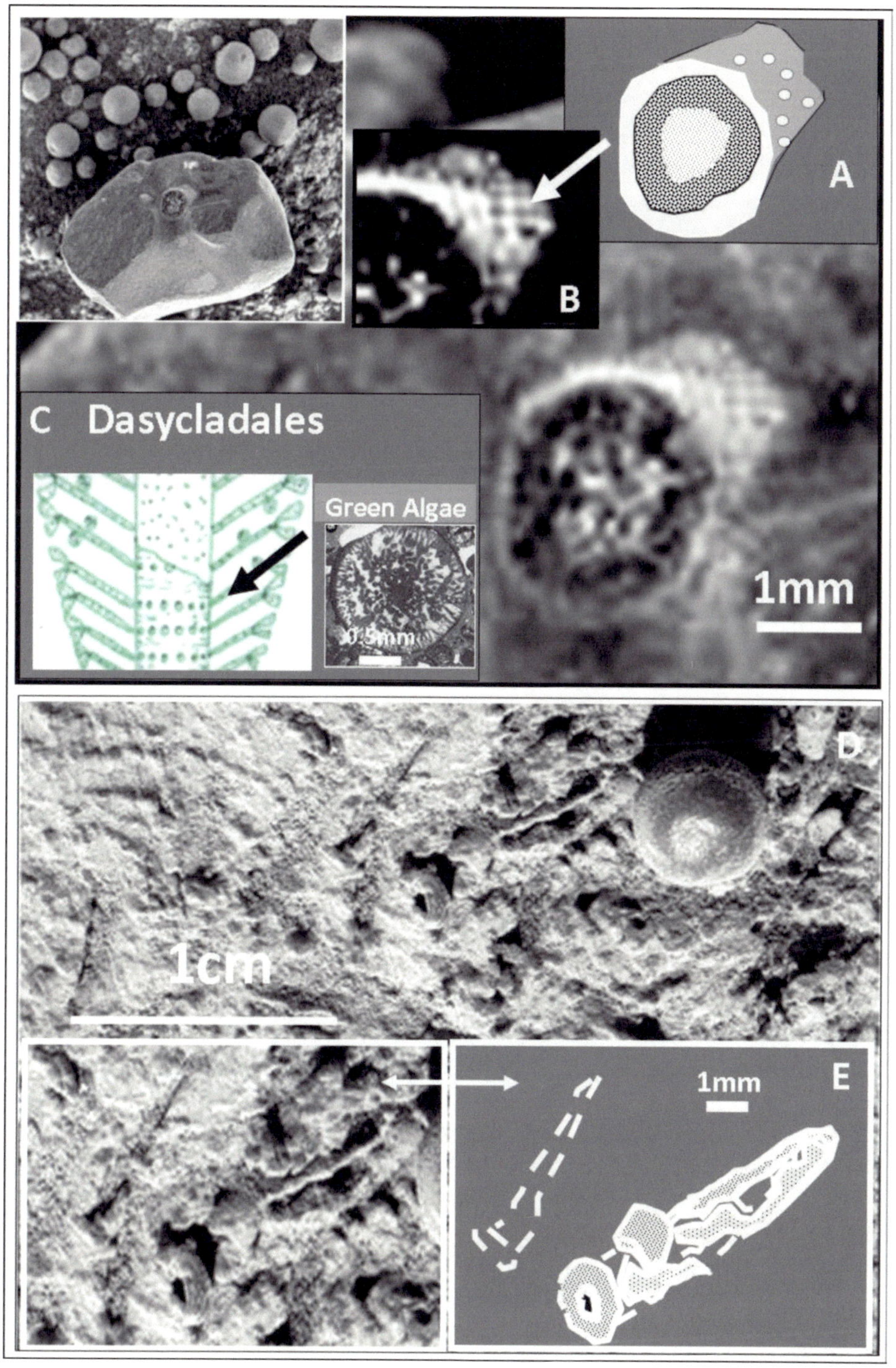


**Figure 10.**
*Different conical bodies, detected at Meridiani Planum by Opportunity; possible biological remnant of fossils "incertae sedis". These microscopic cones have littler bigger dimension. On the top (frames A, B) the cone has similar size and shows an internal zoned structure of a possible algal stem (C), having a number of lateral structures (arrows), resembling (far, not confirmed) terrestrial Dasycladales laterals side. Below (D, E) the cracked cone is littler bigger and show a collar.*

crystal samples. Significant departures from a rectangular distribution, suggesting orientation specificity, were observed in the Mars sample and also by the stromatolite algal filaments, Gymnocodium, Jarosite and feldspar phenocrysts. The percentage

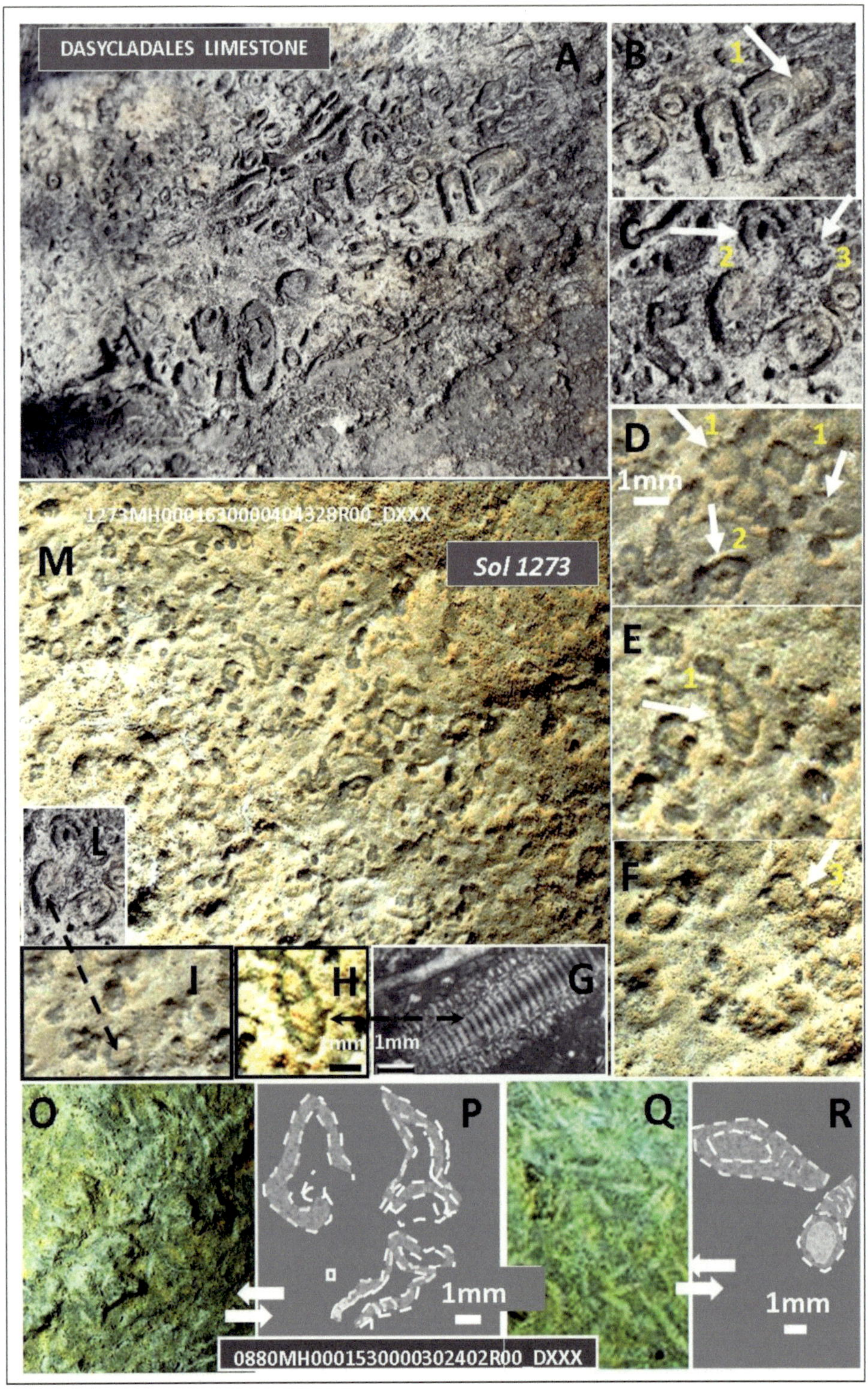


**Figure 11.**
*Comparison of a Dasycladales limestone (b/n pictures) to Martian putative fossils, at Sols 880 (frames O-R) and 1273 (frames D-F, H,I,M). Resembling (far, not confirmed) features of terrestrial Dasycladales include millimetric dimensions, a calcareous (conical or tube-like) stem, occasionally having regular discontinuity (frame G; from [62]), as are undulation or fissures. Such structures are visible on the above b/n pictures, as well on putative Martian fossils where regular transversal lines (arrow 1) or tube-like structures and their transversal sections (arrows 2 and 3) are visible. Careful observation shows knurled shells (for algal laterals). Similar structures appear to be seen in the rover's images (see the arrows), including the knurled shells.*

**Figure 12.**
*Fossil and living terrestrial algae together with mineral crystals used to compare with the "rice grains" in* ***Figure 2****: (1) Population of fossil stromatolite algal filaments (From: [63], their* ***Figure 5****); (2) Population of fossil algae (Gymnocodium group, possibly a red or green alga) (From: [41], Plate 57 7/8); (3) Sample image of living* Euglena viridis*; (4) Gypsum crystals (Gesso-2), (5) Jarosite crystals (jarosite 4138d, www.dakotamatrix.com), (6) crystals of feldspar (phenocrysts) (1200px-Montblanc-granite-phenocrysts. Mineral from Mineralienatlas.de., authorized).*

of profiles exhibiting a fusiform shape or a degree of curvature also varied among images being generally greatest for the Mars sample and the terrestrial microalgae and least for the mineral deposits.

A PCA of the data resulted in the extraction of two Principal Components (PC's) accounting for 96% of the total variance (PC1 = 77%, PC2 = 19%) indicating that separation of images along PC1 is more significant than along PC2. A plot of the 10 images in relation to PC1 and PC2 is shown in **Figure 13** with significant correlations between the factor loadings of the images on the PC and the various metrics. **Figure 13** shows: (1) that the three images of gypsum crystals (G1–3) have very similar metrics and form a distinct cluster not closely related to the Mars sample or to any of the investigated algae, (2) neither feldspar phenocrysts nor jarosite crystals are closely related to the Mars sample (3) Gymnocodium (Gym) and the Stromatolite Algal filaments (SA) are more closely related to the gypsum crystals in their metrics than Euglena (Eug) and (4) of the studied microalgae. Correlations between factor loadings and the various metrics suggest, the proportion of profiles with a fusiform shape, the proportion of curved profiles and the degree of variation in profile widths are all significantly correlated with PC1 and PC2 and, therefore, are the most important of the metrics distinguishing among images, the Martian sample and terrestrial Euglena displaying the most consistent widths and having a greater proportion of fusiform and curved profiles than the mineral deposits and other terrestrial microalgae.

| Image | N | V\M | CV-L | CV-W | L/W | Fit to LN | Fit to RD | % FF | % Curved |
|---|---|---|---|---|---|---|---|---|---|
| | *Metrics* | | | | | | | | |
| Mars | 411 | 0.78 | 42% | 16% | 4.1 | 0.03 | 0.09** | 64% | 36% |
| SA | 79 | 0.95 | 95% | 25% | 6.5 | 0.08 | 0.15* | 51% | 29% |
| Gym | 162 | 0.48 | 104% | 61% | 3.1 | 0.05 | 0.30** | 64% | 32% |
| Eug | 111 | 0.32 | 41% | 16% | 4.4 | 0.04 | 0.09 | 86% | 71% |
| Gyp-1 | 169 | 0.35 | 74% | 45% | 2.3 | 0.01 | 0.07 | 105% | 1% |
| Gyp-2 | 159 | 0.46 | 73% | 63% | 1.7 | 0.04 | 0.07 | 17% | 0% |
| Gyp-3 | 147 | 0.53 | 59% | 37% | 2.0 | 0.04 | 0.09 | 10% | 0% |
| Jar-1 | 128 | 1.08 | 34% | 44% | 3.7 | 0.03 | 0.19** | 18% | 12% |
| Jar-2 | 76 | 0.48 | 37% | 34% | 3.6 | 0.05 | 0.11 | 15% | 10% |
| FD | 159 | 0.65 | 47% | 47% | 3.0 | 0.04 | 0.33** | 1% | 1% |

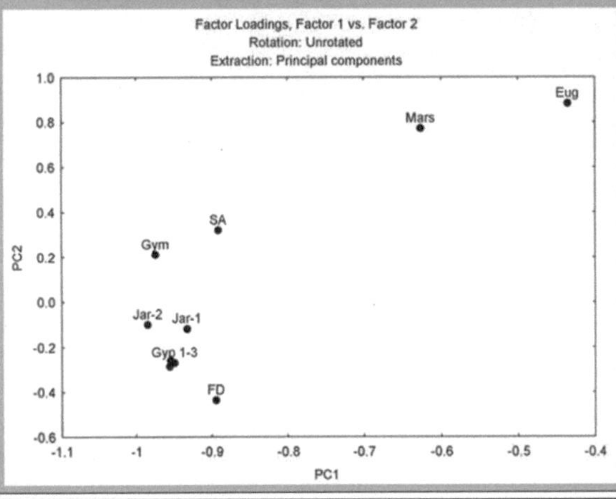


A– *Euclidean Morphometry*

| *Metrics* | *PC1* | *PC2* |
|---|---|---|
| V/M | -0.13 | 0.03 |
| CV (lengths) | -0.38 | -0.07 |
| CV (widths) | -0.73* | -0.72* |
| L/W | 0.39 | 0.62 |
| Fit to LN | -0.09 | 0.20 |
| Fit to RD | -0.26 | -0.22 |
| % Fusiform | 0.74* | 0.95* |
| % Fus. curved | 0.84* | 0.94* |

B– *Fractal Morphometry*

| *Metrics* | | | | |
|---|---|---|---|---|
| *Analyzed structure* | *N* | *Mean* | *SD* | |
| Mars | 30 | 1.570 | 0.047 | |
| Gypsum | 10 | 1.615 | 0.034 | P < 0.01 |
| E. viridis | 30 | 1.581 | 0.055 | P = n.s. |

**Figure 13.**
*Results of Euclidean morphometric investigation on "rice grains". On the top (A) the plot shows results of Euclidean morphometry analysis, differences and similarities, between the 10 images analyzed: PC1 against PC2 (Eug = Euglena, Gyp 1–3 Images of gypsum crystals, FD = Feldspar phenocrysts, Gym = Gymnocodium, Jar1 - Jar2 = Jarosite crystals, SA = Stromatolite algal filaments). Below (B), comparison of fractal dimension values (Mars, "rice grains" vs. Gypsum (P < 0.01) and vs. Euglena viridis (P = n.s.)). Its fractal dimension permit to distinguish them from the mineral negative control, while it is not possible to distinguish the Martian features from the biologic control, perfectly superimposable among them.*

### 5.2.2 Fractal morphometry

Fractal analysis data are summarized in **Figure 13**, bottom, B. Fractal dimension of the Martian "rice grains" is lower than the one of the negative control, gypsum, with high statistical significance ($p < 0.01$). Vice versa, fractal dimension of the Martian "rice grains" overlaps the one of the unicellular alga *Euglena viridis*, positive control (D = 1.570 + 0.047 vs. 1.581 + 0.055, mean + SD, n = 30 per each sample).

## 6. Discussion

The first presence of organic matter on the Red Planet was revealed, even if initially misunderstood, by the Viking's pyrolysis gas chromatography–mass spectrometry (GC–MS) analysis of Martian regolith [64]. More recently, the presence of chloromethane and dichloromethane, as markers of organic matter on Mars, were confirmed by SAM on Curiosity [65]. Since the Viking landers, organic matter has been repeatedly detected in Martian meteorites [66]. Hence, Curiosity rover drilled into the three-billion-year-old mudstones from four areas in Gale Crater and especially at Mojave and Cumberland sites, revealing many organic compounds, including thiophenic, aromatic, and aliphatic compounds, that were released at temperatures from 500° to 820°C [67], and reported in **Table 1**. The variety of different carbon-containing compounds provides evidence of possible macromolecules in the Martian regolith. Interpreting their presence, we can recall the kerogens observed on Mars: they are a type of organic molecule that can be easily associated with life (stromatolites), but, viceversa, it is also present in carbon-rich meteorites, in interplanetary dust particles, and in igneous rocks, where life is not present. Nevertheless, the thiophenes observed on Mars, should be strongly suggestive of life [68], being easily explained as a result of biologically related sulfur incorporation into organic matter during early diagenesis.

Moreover, it has been suggested that biominerals could be important indicators of life and thus could play an important role in the search for past or present life on Mars as on Earth [58, 69, 70]. Organic components themselves (Kerogene and Tiophene) are often associated with biominerals and are believed to play crucial roles in both pre-biotic and biotic reactions. They have been found in the fossil record that date back to the Precambrian and were used on Earth as evidence of the biogenicity of Archean stromatolites [56].

On Mars, and in particular at Mojave targets, morphological observations of dendritic, nodular and laminated, harder structures (and complex organics, as well biominerals occurrence), may suggest a common origin, and may represent possible developmental stages of a single entity. Their variable dimensions, scattered distribution, and uncommon shape, show the same morphology as terrestrial microbialites. In this frame, noteworthy are the small nodular and encrusting microbialites, which are found in a wide range of lacustrine environments and in thin laminated mudstone, and they have been attributed to moderate wave agitation [71]; convincing parallels being visible, in the lake stromatolites of the West Germany lower Permian (Lauterecken Formation); as well as examples of stromatolites in nodular settings, forming larger cemented complexes known in current alkaline (pH > 9) fresh lakes (Salda Lake, Turkey).

In this context, the irregular shapes assumed by the harder structures containing complex organics and biominerals most likely represent a results of bacterial or microalgae extracellular polymeric substances, according to an organic mineralization process present during diagenesis.

The spatial development of stromatolites is important in interpreting their eventual structures. Hence, the basic structure of microbialitic sediments are essentially laminar (in plane), nodular (balls or lumps) and/or elongated (linear). These structures can also merge, respectively resulting in stromatolites, thrombolites, dendrolites and with ever-larger combinations providing all the typical known morphologies. The observed structures and morphologies, shown in **Figures 4–11** and all of those described to date in various studies, are all typical of microbialitic world.

In general, the complexity and distinctiveness of biological structures increase with size and degree of biological evolution. There is still controversy on Earth regarding the biogenicity of some primordial microscopic structures and specialists attempt to solve these problems using instrumental insights and further laboratory investigations. These problems are generally related to the presence of possible very ancient microbial structures, having micrometric or sub-micrometric dimensions.

Indeed, microfossils, with a size of hundreds of microns, are more complex and distinctive, and on Earth other investigations are usually not necessary to recognize their biogenic nature.

In terms of "relevance for morphological recognition of biogenic structures", three domains could be distinguished: microbes, microfossils and fossils. Although doubts have often been expressed about the visual unambiguity of Martian microstructures, the described morphologies here described and related to putative "microalgae" should be considered unambiguous. In fact, the Mahli images that we have analyzed most frequently have pixels in the range 20–30 microns. For example, **Figure 9** (taken at Sol 880) which contain the lozenge-shaped bodies and the "cornucopia", have a pixel dimension of about 25 microns whereas the analyzed objects have millimetric or submillimetric dimensions and contain hundreds of colored pixels. As consequence, a single septate partition of the elongated structure shown in **Figure 9** (frame 1), having a dimension of about 0.1 mm, contains more than 16 colored pixels, which enables the septate structures to be unambigously observed.

Such filamentous segmented structures, having a cross section of 0.09–0.30 mm, are common in Martian sediments and have been described in previous papers [16, 72, 73]. In general, we can state that septate filament-like structures are frequent in terrestrial algal-like biota and that the Martian structures, that we have highlighted, are morphologically similar to a wide range of terrestrial counterparts (i.e., Epimastopora green alga; **Figure 14**).

The fossil record of septate bodies and the filaments are abundant, and are mainly characteristic of three big groups, i.e. Oscillatoriopsis, Megathrix, and those phosphatized tubular fossils (**Figure 15**) in the Ediacaran Weng'an Biota [74]. Oscillatoriopsis is characterized by unbranched, unsheathed, uniserate cellular trichome. The cells are uniform in length and diameter within the same trichome with no constrictions at the cell boundaries. Butterfield [75] recognized four species of Oscillatoriopsis according to their diameter, i.e., O. vermiformis Schopf, 1968, 1–3 μm; *O. obtusa* Schopf, 1968, 3–8 μm; O. amadeus Schopf and Blacic, 1971, 8–14 μm and O. longa Timofeev and Hermann, 1979, 14–25 μm.

Megathrix, however, are tubular microfossils typically less than 100 μm wide and several hundred μm long. These tubes, rarely branched, are characterized by evenly spaced transverse cross-walls which are complete or incomplete. Complete cross-walls are corrugated or flat and most are regularly intercalated with incomplete cross-walls. Incomplete cross-walls are flatter or less strongly corrugated than the complete examples and they have central perforations typically of similar size within the same specimen, although the perforation size may vary between specimens. Liu et al. [74] described five species of tubular microfossils from the Ediacaran Doushantuo Formation at Weng'an, Guizhou Province, South China. They also have complete and incomplete cross-walls [76]. However, the diameter of the Doushantuo species (mostly 100 μm – 250 μm in diameter) is much greater than Megathrix longus Yin L. and they all have flat rather than corrugated cross-walls. Of the five Doushantuo species, Ramitubus increscens Liu P. [74] and *Ramitubus decrescens* Liu P. [74] are both characterized by regularly dichotomous branching and rare incomplete cross-walls and by tetragonal tubes while *Crassitubus costatus* Liu P. [74] by a curved cylindrical tube with a longitudinal ridge, while *Quadratitubus orbigoniatus* Xue Y. [77] is characterized by a ridge. Finally, *Sinocyclocyclicus guizhouensis* Xue Y. [77] is most similar to *Megathrix longus* Yin L. except the former has greater diameter and flat cross-walls [76, 77]. We cannot be certain whether the septate filaments have corrugated or flat cross-walls but some ring-shaped (R) in **Figures 8** and **9** may represent central perforations on the cross walls; and very curved (VC) in the same figure have short incomplete walls, all supporting a resemblance to *Megathrix longus* Yin L.. The cornucopia-like

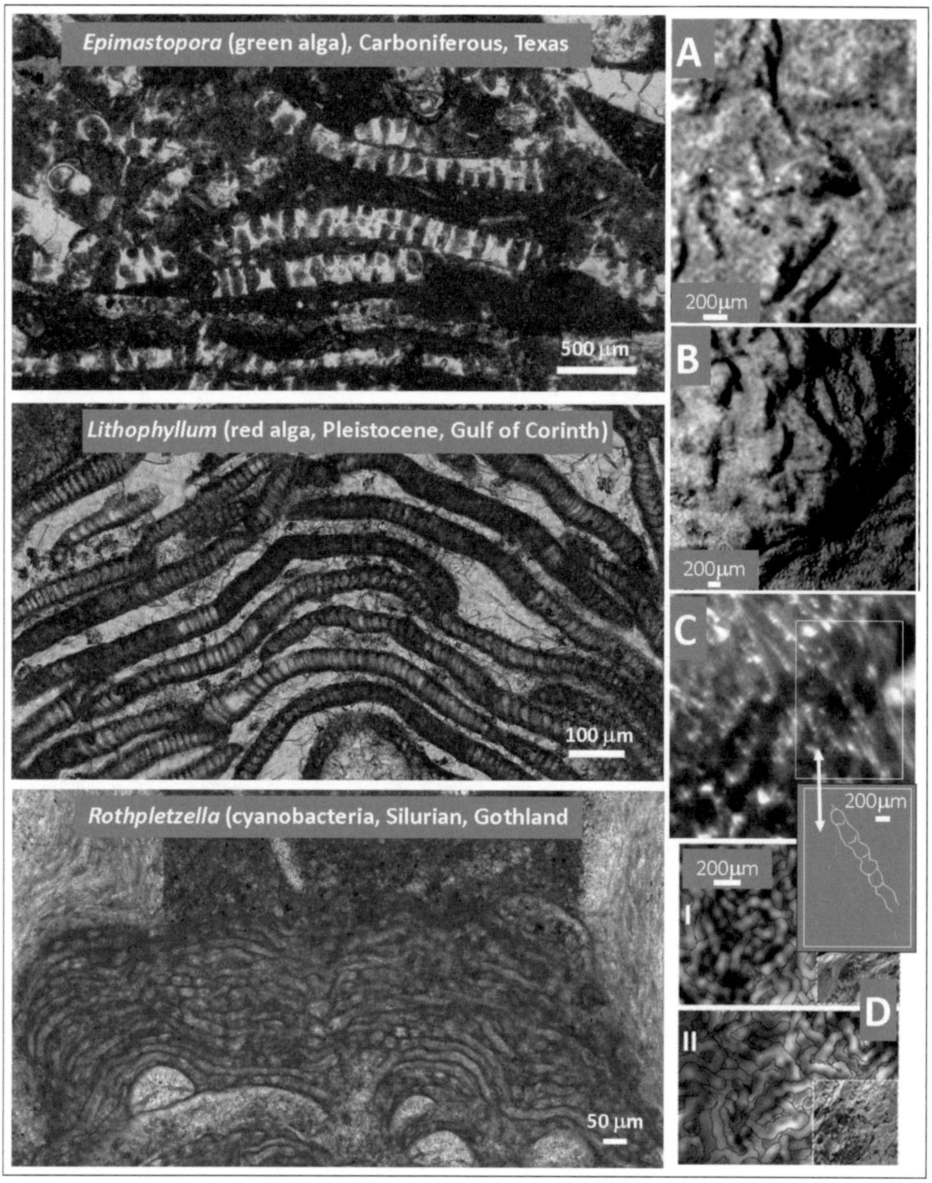


**Figure 14.**
*Septate filament-like structures are biological "in principle" on Earth given a large number of algal structures having similar shape but different dimensions. On the left, three examples of different biota. Frames on the right represent Martian samples of similar structures. Right on the top (A), an enlarged cutting of Sol 880 image showing hard filament, in relief and septate. Frame B represent a reproduction of filamentous structure described on **Figure 6**, on frames H and I. Frame C is a ChemCam (CR0_631854670PRC_F0781160CCAM03640L1) image catting showing some septate filaments, resembling intertwined filaments of spherules. In D similar structures observed in previous works and compared to some stromatolite similar structures.*

structures may represent oblique sections of tubular fossils with septate filaments such as present in *Megathrix longus* Yin L. or *Sinocyclocyclicus guizhouensis* Xue Y. [77] with possibly cyanobacteria affinity (**Figure 15**).

Moreover, morphometric investigations suggest a PCA and fractal dimensions of the "rice grains" with an affinity far from the mineral deposits studied such as gypsum, jarosite, and feldspar phenocrysts. In effect, phenocrysts are euhedral and angular whereas many of the Martian deposits are fusiform and exhibit a degree

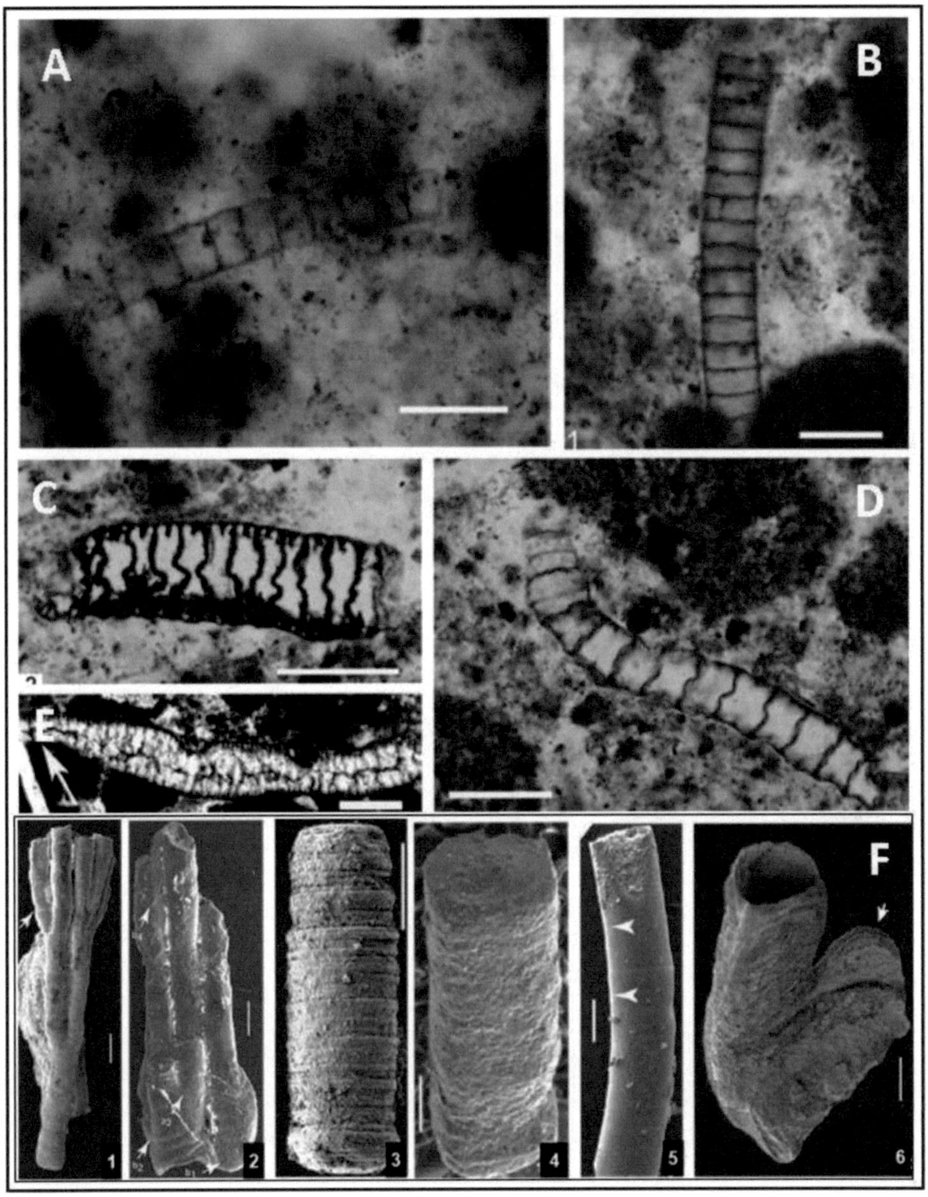


**Figure 15.**
*Microphotographs of a set of tubular septate-likes bodies thin sections, comparable to Martian samples. Frames A and B:* Oscillatoriopsis longa *Timofeev and Hermann, 1979; scale bars are 50 μm. Frames C-E:* Megathrix longus *Yin L. from the lower Yurtus and lower Yanjiahe formations; scale bars are 100 μm. Frame F: 1.* Ramitubus increscens *Liu P., 2008 (scale bar is 200 μm); 2.* Ramitubus decrescens *Liu P., 2008 (scale bar is 200 μm); 3.* Sinocyclocyclicus guizhouensis *Xue Y., 1992 (scale bar is 100 μm); 4.* Quadratitubus orbigoniatus *Xue Y., 1992 (scale bar is 100 μm); 5.* Crassitubus costatus *Liu P., 2008 (scale bar is 100 μm); 6.* Yangtzitubus semiteres *Liu P., 2008 (This one is silicified and also from Ediacaran Doishantuo Formation; scale bar is 50 μm).*

of curvature. However, some gypsum and jarosite crystals exhibit a more fusiform shape but only a small proportion exhibited a degree of curvature which itself is regarded as a microbial biosignature [73].

Morphometric comparisons of "rice grains" with Euglena and Dasycladales have also been investigated, due to their morphological affinity. The fossil record of Euglena, however is rare [78–80] and only a fossil, called Moyenia, has been recorded from Late Ordovician non-marine deposits. This record was suggested by

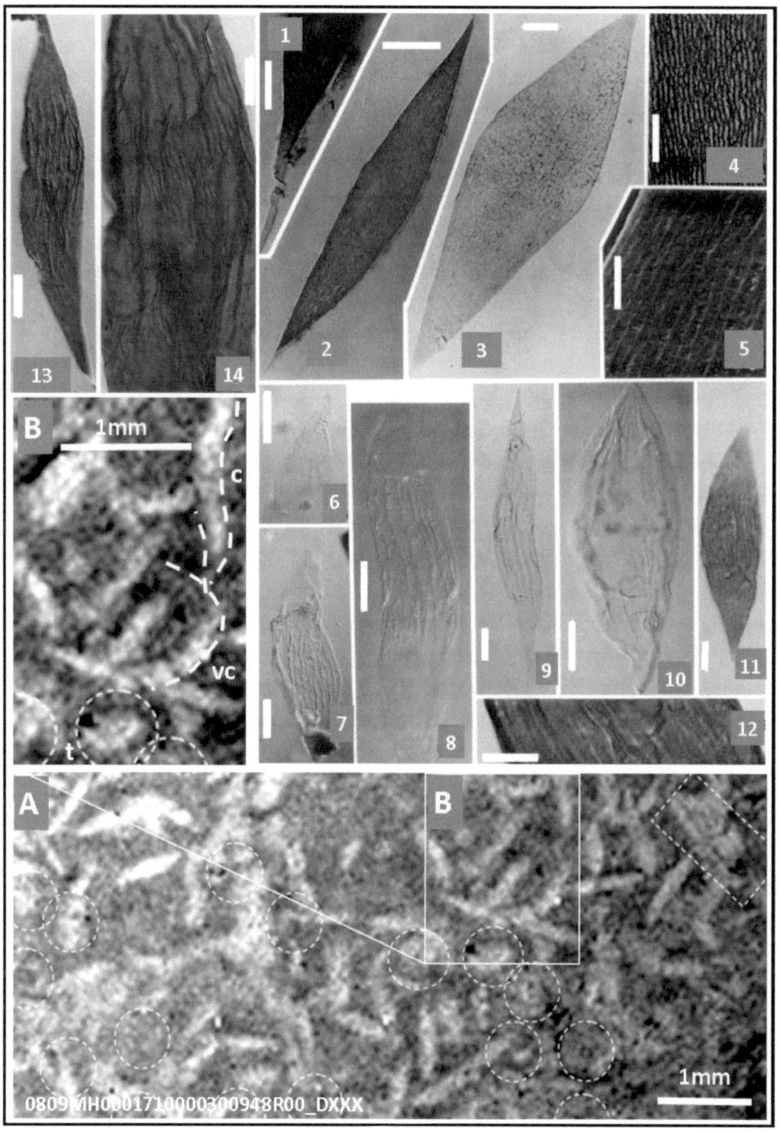


**Figure 16.**
*Structural features of Euglenae (from Loeblich and Tappan [81]) in comparison to the "rice grain" (A,B). "Rice grain" show curved (c) and very curved (vc) structures with holed transversal sections (t). 1–5 -* Cleithronetrum cancellatum *Loeblich and Tappan, 1979 1, showing smooth distal portion of process, Scale bar 10 µm; 2, same specimen, showing slightly asymmetrical fusiform shape, Scale bar 30 µm; 3, nearly symmetrical lozenge-shaped outline, Scale bar 20 µm; 4, enlargement of surface of specimen of* ***Figure 2****, showing longitudinal sinuous and anastomosing ridges that bear tiny grana, Scale bar 10 µm; 5; scanning electron microscope photograph, showing partial or complete bridges between and perpendicular to longitudinal ridges and deep pits separating the bridges, Scale bar 5 µm. All from the Mountain Lake Member of the Bromide Formation of Oklahoma. 6,7 -* Eupoikilofusa striata *Staplin, Jansonius and Pocock n. comb. 6, holotype, blunt-tipped polar process that lacks the ridges found on median area of vesicle, Scale bar 20 µm; 7, overall view of holotype, showing longitudinal ridges, Scale bar 20 µm. From the Middle Ordovician, Trenton Formation, Anticosti Island, Canada. 8,9 -* Eupoikilofusa ctenista *Loeblich and Tappan, 1979. 8, holotype, enlargement to show nature and distribution of ridges on vesicle, Scale bar 20 µm; 9, lesser magnification of same, showing pointed processes that lack the ridges found on the vesicle, Scale bar 20 µm. From the Sylvan Shale of Oklahoma,* Eupoikilofusa platynetrella *Loeblich and Tappan, 1979, holotype, showing slightly asymmetrical vesicle and distribution of vesicle ribs, Scale bar 20 µm. From the Sylvan Shale of Oklahoma. 11, 12-*Eupoikilofusa anolota *Loeblich and Tappan, 1979. 11, holotype, showing relatively broad and low discontinuous ridges, Scale bar 20 µm; 12, enlargement of surface to show wall sculpture, Scale bar 10 µm. From the Sylvan Shale of Oklahoma, 13,14 -* Eupoikilofusa parvuligranosa *Loeblich and Tappan, 1979. 13, holotype, nearly symmetrical fusiform vesicle with ridges that die out toward polar processes, Scale bar 20 µm; 14, enlargement of holotype to show sinuous ridges and grana aligned in rows on vesicle, Scale bar 10 µm. From the Sylvan Shale of Oklahoma.*

Colbath & Grenfell to be a possible fossil pellicle (cell wall) of a euglenid, based on their surface morphology, whereby the spiral pattern of ridges on the pellicle resembles that of some photoautotrophic euglenids in Monomorphina Mereschkowsky, 1877 [81]. **Figure 6** is particularly interesting and the cone present could easily be accepted as a fossil but given its incomplete preservation, it could be similar to a portion of Cloudina, a small shell fossil, or an example of other conotubular fossils (**Figure 16**). We emphasise, however, that the degree of variation in profile, width and shape and the degree of curvature of the profiles, both criteria highlighted by Williams [73] with consistency in width and curvature, are indicative of biogenicity and fractal analysis clearly confirms that the "rice grains" cannot be identified with mineral abiological structure as the gypsum, a result with a high statistical significance ($p < 0.01$).

Nevertheless, our conclusion in this paper concerning the analyzed Martian microstructures is not that they can be identified with the terrestrial species described but that the characteristics of shape and degree of complexity, make it probable that on Mars, 2.2 billion years ago, there were complex life forms, analogous to terrestrial eukaryotic cells. From a biological point of view, it is not unlikely that similar forms of life, with so many structural similarities, have developed independently on two different planets.

Results presented in this article can easily be interpreted as a phenomenon of evolutionary convergence, a phenomenon which is extremely widespread in terrestrial life forms. We can recall mammals and octopuses having camera-like eyes with an iris, a lens and a retina or the wings of bats and birds or the shape of sharks and dolphins: analogous environments producing the "same" shapes and structures without any evolutionary linkage. There no any problem for what concerns the time and the environment, on Earth, and, respectively, on Mars. Age of 2.2 billion years ago on Mars it is the same age in which complex, eukaryotic, cells, appeared on Earth, so there is no any problem for the time that could be need on Mars to produce those type of cells. No any problem, also, for the environment: the Earth aging 2.2–2.5 Gy ago is the one at the time of the oxygen crisis and of the "snowball Earth" [82], an anoxic and cold Earth as Mars of that age.

## 7. Further experimental observations for compelling evidence of life on Mars

Will never be able to definitively prove the existence of life on the Red Planet? The search for definitively proving the presence of life on Mars is one of the outstanding scientific challenges of our time [69, 83, 84]. We have described in this paper how the Curiosity landed region was clearly demonstrated as a fluvial-deltaic-lacustrine environment [2]. There are strong evidences that the whole surface of early Mars was habitable and several biomarkers were found (including microbialite/stromatolite-like structures and *bona fide* microfossils). Orbital and rover data reveal fluvial valley networks, paleolake deposits, alluvial fans/deltas entering these lakes, clearly revealing sustained precipitation during the Noachian, Hesperian and Amazonian periods, since 4.5 to 0.9 Gy ago, when an extensive glacial event resurfaced the paleolakes themselves [85, 86].

Nevertheless we need complex approaches for conclusively establish the presence of life [58] (NASA MEPAG's program at http://mepag.jpl.nasa.gov/reports/index.html), and, in particular, when tests need to be performed by rovers or landers in a planet hundred million kilometers far from Earth. E.g., we can recall the lot of papers supporting the presence of stromatolites on Mars by morphological approaches, as also here discussed.

The meaning of the possible presence of stromatolites on Mars is enormous. Stromatolites result from the activity of different microbial communities and not the product of a single microorganism, suggesting a real ecosystem on the planet if the presence of stromatolites should be definitively proved. But the morphology of stromatolites as indicator of biogenicity may be ambiguous, similar shapes can be produced by both abiotic and biotic processes [87]. In effect, geologists on Earth needs to study macro-, micro- and ultramicroscopic details of the putative stromatolites, as well as their geochemistry, Raman spectroscopy of their constituents, additional chemical, mineralogical (e.g. magnetite and pirite) and isotopic ratios analysis of redox-sensitive heavy elements (mainly of C, Fe, S, N, Mo, Cu, U, Ce), together which the sedimentological, stratigraphic and palaeoenvironmental context examined [69, 88, 89], to arrive at a reasonable conclusion about the biotic or abiotic origin of the hypothesized biostructures. Much more, the presence of carbonaceous matter (e.g., kerogen, bitumen, molecular biomarkers) in the same area needs to be evaluated, together which their isotopic compositions, drilling meters below the surface in order to collect organics that had no destroyed by the UV flux of the Sun in the present-day Mars. Analysis that involve the use of traditional tested technology, as well new experimental and miniaturized biosensors for "in situ" testing. Not a simple work, so it's not strange that till today it was not possible to reach unanimous agreement among astrobiologists concerning the presence of life on the Red Planet. However, we feel that the next Mars missions by NASA and ESA, together which Mars Sample Return missions, should be able to reach the goal.

## 8. Conclusion

On Mars, at Gale Crater, a past environment favourable to life and for a broad span of geological time, has been discovered by various authors [1, 2, 49], as well as the occurrence of many micro, meso and macrostructures similar to terrestrial stromatolites, microbialites and algae [18, 50]. All these items are suggestive of possible biological parallels between Earth and Mars. In the present paper, we show morphological and morphometric analyses of the whitish millimetric shapes (we nicknamed "rice grains"), detected by the rover Curiosity at sols 809 and 880 (Mojave target) in the lacustrine Murray Formation (attributed to pseudomorph crystals of sulfate by previous Authors, [7]). Specimen which are incompatibile to Gypsum, Jarosite, or Feldspar crystals, but show a high shape affinity to life forms such as the Euglenoids. Hence, the microstructures investigated in this study, together with chemical and mineralogical converging data of the outcropping where they are embedded, suggest the possible existence of microbial, and/or little more complex life forms, in the past history of Mars.

## Acknowledgements

We thank Filippo Barattolo, Professor at University of Naples (IT) and Joan Bucur, Professor at Department of Geology of Babes-Bolyai University (RO) for the support given in the analysis of algal-like biota and in excluding Dasycladales attribution. We are particularly grateful to Prof. Munneke, Professor at Friederich-Alexander University of North Bayern, for providing us the images of the terrestrial septate filaments shown on **Figure 15**. This work would not have been possible without NASA"s images and data availability, for which we are grateful.

## Author details

Vincenzo Rizzo[1*], Richard Armstrong[2], Hong Hua[3], Nicola Cantasano[4], Tommaso Nicolò[5] and Giorgio Bianciardi[5]

1 Emeritus, Consiglio Nazionale delle Ricerche-ISP, Messina, Italy

2 Aston University, Birmingham, UK

3 State Key Laboratory of Continental Dynamics, Department of Geology, Early Life Institute, Northwest University, Xi'an, China

4 Italian National Research, I.S.A.FO.M. U.O.S., Cosenza, Italy

5 University of Siena, DSMCN, Siena, Italy

*Address all correspondence to: enzo45.rizzo@gmail.com

## References

[1] Grotzinger, J.P., D.Y. Sumner, D.Y, Kah, C, Stack, K., Gupta, S, Edgar, L., Rubin, K. Lewis, J., Schieber, N., Mangold, R., Milliken, P.G., Conrad, D., DesMarais, J., Farmer, K., Siebach, F., Calef III, Hurowitz, J., McLennan, S.M., Ming, D. Vaniman, J., Crisp, A., Vasavada, A., Edgett, K.S., Malin, D., Blake, D., Gellert, R., Mahaffy, P., Wiens, R.C., Maurice, S., Grant, J. A., Wilson, S., Anderson, S., Beegle, L., Arvidson, R., Hallet, B., Sletten, R.S., Rice, M., Bell III, J., Griffes, J., Ehlmann, B., Anderson, R.B., Bristow, T.F., Dietrich, W.E., Dromart, G., Eigenbrode, J., Fraeman, A., Hardgrove, C., Herkenhoff, K., Jandura, L., Kocurek, G., Lee, S., Leshin, L.A., Leveille, R., Limonadi, D., Maki, J., McCloskey, S., Meyer, M., Minitti, M., Newsom, H., Oehler, D., Okon, A., Palucis, M., Parker, T., Rowland, S., Schmidt, M., Squyres, S., Steele, A., Stolper, E., Summons, R., Treiman, A., Williams, R., Yingst A., MSL Science Team, 2014. A habitable fluvio-lacustrine environment at Yellowknife Bay, Gale Crater, Mars. Science 343, 6169, 1242777.

[2] Grotzinger J.P., Gupta S., Malin M.C. et al., 2015. Deposition, Exhumation, and paleoclimate of an ancient lake deposit, Gale Crater, Mars. Science 350 (6257), 1-12.

[3] Halliday A.N., Wänke H., Brick J.-L. e R. N. Clayton. The Accretion, Composition and Early Differentiation of Mars. In Space Science Reviews, vol. 96, n. 1/4, 2001, pp. 197-230.

[4] Bibring, J. P. et al., 2006. Global Mineralogical and Aqueous Mars History Derived from OMEGA/Mars Express Data. Science, vol. 312, pp. 400-404, DOI:10.1126/science.1122659

[5] Ashley Y., 2008. Impact May Have Transformed Mars. Science News org.

[6] Head, J.W., Wilson, L., 2001. The Noachian-Hesperian Transition on Mars: Geological Evidence for a Punctuated Phase of Global Volcanism as a Key Driver in Climate and Atmospheric Evolution. 42nd Lunar and Planetary Science Conference.

[7] Martin, P.E., Farley, K.A., Baker, M.B., Malespin, C.A., Schwenzer, S.P., Cohen, B.A., Mahaffy, P.R., McAdam, A.C., Ming, D.W., Vasconcelos, P.M., Navarro-González, R.A., 2017. Two-Step K-Ar Experiment on Mars: Dating the Diagenetic Formation of Jarosite from Amazonian Groundwaters. Journal of Geophysical Research: Planets 122 (12), 2803-2818. https://doi.org/10.1002/2017JE005445

[8] Kostama, L., Kreslavsky, H., 2006. "Recent high-latitude icy mantle in the northern plains of Mars". Agu.org.

[9] NASA, 2007. Mars' South Pole Ice Deep and Wide". Nasa reports, on jpl.nasa.gov.

[10] Weber, K.A., Trisha, L., Spanbauer, M., Wacey, D., Kilburn, M.R., Loope, D.B., Kettler, R. M., 2012. Biosignatures link microorganisms to iron mineralization in a paleoaquifer. Geology 40, 747-750.

[11] Westall, F., Foucher, F., Bost, N., Bertrand, M., Loizeau, D., Vago, J.L., Kminek, G., Gaboyer, F., Campbell, K.A., Bréhéret, J.G., Gautret, P., Cockell, C.S., 2015. Biosignatures on Mars: What, Where, and How? Implications for the search of Martian life. Astrobiology 15 (11), 1-33.

[12] Bianciardi, G., Rizzo, V., Cantasano, N., 2014. Opportunity Rover's image analysis: microbialites on Mars? International Journal of Aeronautical and Space Sciences 15, 419-433.

[13] Bianciardi, G., Rizzo, V., Maria E. Farias, Cantasano, N., 2015.

Microbialites at Gusev crater, Mars. Astrobiol Outreach 3(5), http://dx.doi.org/10.4172/2332-2519.1000143

[14] Noffke, N., 2015. Ancient sedimentary structures in the < 3.7b Ga Gillespie Lake member, Mars, that compare in macroscopic morphology, spatial associations, and temporal succession with terrestrial microbialites. Astrobiology 15, 1-24.

[15] Rizzo, V., Cantasano, N., 2009. Possible organosedimentary structures on Mars. International Journal of Astrobiology 8, 267-280.

[16] Rizzo, V., Farias M. E., Cantasano, N., Billi, D., Contreras, M., Pontenani, F., Bianciardi, G., 2015. Structures/Textures of living/fossil microbialites and their implication in biogenicity. An astrobiological point of view. ACB 4 (3), 65-82.

[17] Rizzo, V., Cantasano, N., 2017. Structural parallels between terrestrial microbialites and Martian sediments: are all cases of "Pareidolia"? International Journal of Astrobiology 16, 297-316.

[18] Rizzo, V., 2020. Why should geological criteria used on Earth not be valid also for Mars? Evidence of possible microbialites and algae in extinct Martian lakes. International Journal of Astrobiology, 1-12.

[19] McKay, C.P., 1997. The search for life on Mars. Origins Life Evolution Biosphere 27, 263-289.

[20] Carr, M.H., 1981. The surface of Mars. Yale University Press, New Haven, CT.

[21] McEwen, A.S., Hansen, C.J., Delamere, W.A., Eliason, E.M., Herkenhoff, K.E., Keszthelyi, L., Gulick, V.C., Kirk, R.L., Mellon, M.T., Grant, J.A., Thomas, N., Weitz, C.M.,Squyres, S.W., Bridges, N.T., Murchie, S.L., Seelos, F., Seelos, K., Okubo, C.H., Milazzo, M.P., Tornabene, L.L., Jaeger, W.L., Byrne, S., Russel, P.S., Griffes, J.L., Martinez-Alonso, S., Davatzes, A., Chuang, F.C., Thomson, B.J., Fishbaugh, K.E., Dundas, C.M., Kolb, K.J., Banks, M.E., Wray, J.J., 2007. A closer look at water-related geologic activity on Mars. Science, 317, 1706-1709.

[22] Coates, J.D., Michaelidou, U., O'Connor, S.M., Bruce, R.A., Achenbach, L.A., 2000. The diverse microbiology of (per)chlorate reduction. In: Urbansky, E.D., (Ed.). Perchlorate in Environment, Kluwer Academic/Plenum, New York, 257-270.

[23] Coates, J.D., Achenbach, L.A., 2004. Microbial perchlorate reduction: rocket-fueled metabolism. Nat. Rev. Microbiol., 27, 569-580.

[24] Logan, B., 1998. A review of chlorate and perchlorate-respiring microorganisms. Bioremediat. J., 2, 69-79.

[25] McKay, C.P., Carol R. Stoker, Brian J. Glass, Arwen I. Davé, Alfonso F. Davila, Jennifer L. Heldmann, Margarita M. Marinova, Alberto G. Fairen, Richard C. Quinn, Kris A. Zacny, Gale Paulsen, Peter H. Smith, Victor Parro, Dale T. Andersen, Michael H. Hecht, Denis Lacelle, Wayne H. Pollard, 2013. The Icebreaker Life Mission to Mars: A Search for Biomolecular Evidence for Life, Astrobiology, 13 (4), 334-353.

[26] Stoker, C.R., Zent, A., Catling, D.C., Douglas, S., Marshall, J.R., Archer, D., JR., Clark, B., Kounaves, S.P., Lemmon, M.T., Quinn, R., Rennó, N., Smith, P.H., Young, S.M.M., 2010. Habitability of the Phoenix landing site. J. Geophys. Res., 115.

[27] McKay, D.S., Gibson Jr., E. K., Thomas-Keprta, K.L., Vali, H., Romanek, C.S., Clemett, S.J., Chillier, D.F., Maechling, C.R. and Zare, R.N.,

1996. Search for past life on Mars: Possible Relic Biogenic Activity in Martian Meteorite ALH84001. Science, 273, 924-930.

[28] Di Gregorio, B.E., 2010. Martian sheen: life on the rocks, New Scientists, 2747, 40-43.

[29] Hoover, R.B., 2011. Fossils of Cyanobacteria in CL1 Carbonaceous Meteorites. Journal of Cosmology, 13.

[30] Grilli-Caiola, M.G. and Billi, D., 2007. Chroococcidiopsis from Desert to Mars. (Book Series). Cellular Origin, Life in Extreme Habitats and Astrobiology. Algae and Cyanobacteria in Extreme Environments, Vol. 11, pp. 553-568. Kluwer Academic Publishers, Dortrecht.

[31] Grilli-Caiola, M.G., Billi, D., 2011. Effects of nitrogen limitation and starvation on Chroccocidiopsis (Chrococcales). New Phytologist, 133 (4), 563-571.

[32] Fleming, H.C., Wingender, J., 2010. The biofilm matrix. Nature Reviews Microbiology, 8, 623-633.

[33] Chen, M., Schliep, M., Willows, R.D., Cai, Z.L., Neilan, BA., Scheer, H., 2010. A Red-Shifted Chlorophyll. Science, 329, 5997, 1318-1319.

[34] Riding, R., 2011. The nature of Stromatolites: 3,500 Million years of History and a Century of Research. In: Advances in Stromatolite Geobiology, J. Reitner et al., (Eds.), Lecture Notes in Earth Sciences, 131, 29-74.

[35] Cavalier-Smith, T., 2002. The neomuran origin of archeobacteria, the negibacterial root of the universal tree and bacterial megaclassification. Int. J. Syst. Evol. Microbiol., 52, 7-76.

[36] Aubrey, A.D., Parker, E., Chalmers, J.H., Lal, D., Bada, J.L., 2007. Ironstone concretions – analogs to Martian hematite spherules. Geoscience Research Division, Scripps Institution of Oceanography. Lunar and Planetary Science XXXVIII.

[37] Potter, S.L., Chan, M.A., 2011. Iron mass transfer and fluid flow patterns in Jurassic Navajo Sandstone, southern Utah, USA. Geofluids 11, 184-198.

[38] Jennifer, E., Kyle, P., Schroeder, A., 2007. Microbial Silicification in Sinters from Two Terrestrial Hot Springs in the Uzon Caldera, Kamchatka, Russia. Geomicrobiology Journal 24, 627-641. DOI: 10.1080/01490450701672158.

[39] Ruff, S.W., Farmer, J.D., 2016. Silica deposits on Mars with features resembling hot spring biosignatures at El Tatio in Chile. Nat. Commun. 7, 13554 doi: 10.1038/ncomms13554.

[40] Muller, C., Moreau, D. 2008. Methane and formaldehyde: their abundance and sources on the Earth and Mars. International Journal of Astrobiology 7(1), 63.

[41] Flügel, E., 2010. Microfacies of Carbonate Rocks. (2nd Ed, Springer: Heidelberg, 2010).

[42] Armstrong, R.A., Bradwell, T., 2010. The use of lichen growth rings in lichenometry: some preliminary findings. Geografiska Annaler 92A, 141-147.

[43] Girish, V., Vijayalakshmi, A., 2004. Affordable image analysis using NIH Image/Image J. Indian Journal of Cancer 41, 47.

[44] Syed, A., Armstrong, R.A., Smith C.U.M., 2000. Quantification of axonal loss in Alzheimer's disease: an image analysis study. Alzheimer's Reports 3, 19-24.

[45] Armstrong, R.A., 2007. Measuring the spatial arrangement patterns of pathological lesions in histological

sections of brain tissue. Folia Neuropathologica 44, 229-237.

[46] Pollard, J.H., 1979. Numerical and Statistical Techniques. Cambridge University Press, Cambridge.

[47] Hattis, D.B., Burmaster, D.E., 1994. Assessment of variability and uncertainty distributions for practical risk assessments. Risk Analysis 14, 713-730.

[48] Limpert, E., Stahel, W.A., Abbt M., 2001. Log-normal distributions across the sciences: keys and clues. BioScience 51, 341-352.

[49] Webster, C.R., Mahaffy, P.R., Atreya, S.K., Flesch, G.J., Mischira, M.A., Meslin, P., Farkley, K.A., Conrad, P.G., Christensen, L.E., Pavlov, A.A., Martin-Torres, J., Zorzano, M.P., McKonnochie, T.A., Owen, T., Eigenbrode, J.L., Glavin, D.P., Steele, A., Malespin, C.A., Archer Jr., P.D., Sutter, B., Coll, P., Freissinet, C., McKay, C.P., Moores, J.A., Schwenzer, S. P., Bridges, J.C., Navarro-Gonzales, R., Gellert, R., Lennon, M.T., the MSL Science Team, 2015. Mars methane detection and variability at Gale crater. Science 347 (6220), 415-417. DOI:10.1126/science.1261713.

[50] Joseph, R.G., Graham, L., Burkhard Büde, B., Jung, P., Kidron, G.J., Latif, K., Armstrong, R. A., Mansour, H.A., Ray, J.G., Ramos, J.P., Consorti, L., Rizzo, V., Schild R., 2020. Mars: Algae, Lichens, Fossils, Minerals, Microbial Mats, and Stromatolites in Gale Crater. Journal of Astrobiology and Space Science Reviews 3, 40-111, ISSN 2642-228X, DOI: 10.37720/jassr.03082020.

[51] McBride, M.J., Minitti, M.E., Stack, R.A., et al., 2015. Mars Hand Lens Imagery (MAHLI) observations at the Pahrumps Hills field site, Gale Crater. 46th Lunar and Planetary Science Conference, 2855-2856.

[52] Kah, L.C., 2015. Images from Curiosity: A New Look at Mars. Elements 11 (1), 27-32. https://doi.org/10.2113/gselements.11.1.27.

[53] McLennan, S.M., Grotzinger, J.P., Hurowitz, J.A., Tosca, N.J., 2019. The sedimentary cycle on early Mars. Annual Review of Earth and Planetary Sciences 47, 91-118, doi:10.1146/annurev-earth-053018-060332 (2019).

[54] Westall, F., 2013. Nature and analysis of kerogen associated with early archean biosignatures: lessons for Mars. 44th Lunar and Planetary Science Conference, 1346 pdf.

[55] Nachon, M., Clegg, S.M., Mangold, N., Schröder, S., Kah, L.C. et al., 2014. Calcium sulfate veins characterized by ChemCam/Curiosity at Gale Crater, Mars. J. Geophys. Res. Planets 119, 1991-2016.

[56] Schopf, J.W., Kudryavtsev, A.B., Czaja, A.D., Tripathi, A.B., 2007. Stromatolites and microfossils. Precambrian Research 158, 141-155.

[57] Sànchez-Navas, A., Martin-Algarra, M., 2001. Genesis of apatite in phosphate stromatolites. Eur. J. Mineral. 13, 361-376.

[58] Steele, A., Beaty, D.W., Amend, J., Anderson, R., Beegle, L, Benning, L, Bhattacharya, J., Blake, D., Brinckerhoff, W., Biddle, J., Cady, S., Conrad, P., Lindsay, J., Mancinelli, R., Mungas, G., Mustard, J., Oxnevad, K., Toporski, J., Waite, H., 2005. The Astrobiology Field Laboratory. Unpublished white paper, 72 p, posted Dec., 2005 by the Mars Exploration Program Analysis Group (MEPAG) at http://mepag.jpl.nasa.gov/reports/index.html.

[59] Bazylinski, D., Frankel, R.B., Konhauser, K.O., 2007. Modes of Biomineralization of Magnetite by Microbes. Geomicrobiology

Journal, 24(6), 465-475, https://doi.org/10.1080/01490450701572259

[60] Morrison, S.M., Downs, R.T., Blake, D.F., Vaniman, D.T., Ming, D.W. et al., 2018. Crystal chemistry of Martian minerals from Bradbury Landing through Naukluft Plateau, Gale Crater, Mars. Am. Mineral. 103, 857-871.

[61] Rampe, E.B., Ming, D.W., Blake, D.F., Vaniman, D.T., Chipera, S.J., Bristow, T.F., Morris, R. V., Yen, A.S., Morrison, S.M., Grotzinger, J.P., et al., 2017. Mineralogical trends in mudstones from the Murray formation, Gale crater, Mars. Earth and Planetary Science Letters 471, 172-185.

[62] Bucur, I.I., Săsăran, E., Balica, C., Beleş, D. Bruchental, C., Chendeş, C., Chendeş, O., Hosu, A., Lazăr, D.F., Lăpădat, A., Marian, A.V., Mircescu, C., Turi, V., Ungureanu, R., 2010. Mesozoic carbonate deposits from some areas of the Romanian Carpathians – case-studies. Cluj University Press, pp. 198.

[63] Vai, G.B., Lucchi F.R., 1977. Algal crusts, autochthonous and clastic gypsum in a cannibalistic evaporate basin: a case history from the Messinian of Northern Apennines. Sedimentology 24, 211-244.

[64] Navarro-González, R., Vargas, E., de la Rosa, J., Raga, A.C., McKay, C.P., 2010. Reanalysis of the Viking results suggests perchlorate and organics at midlatitudes on Mars. J. Geophys. Res. Solid Earth 115 (E12), E12010.

[65] Ming, D.W., Archer, P.D., Glavin, D.P., Eigenbrode, J.L., Franz, H.B., Sutter, B., Brunner, A. E., Stern, J.C. et al., 2014. Volatile and Organic Compositions of Sedimentary Rocks In Yellowknifeb Bay, Gale Crater, Mars. Science 343 (6169), 1245267.

[66] Steele, A., McCubbin, F.M., Fries, M.D., 2016. The provenance, formation and implications of reduced carbon phases in Martian meteorites. Meteorit. Planet. Sci. 51, 2203-2225.

[67] Eigenbrode, J.L., Summons, R.E., Steele, A. et al., 2018. Organic matter preserved in 3-billion-years- old mudstones at gale Crater, Mars. Science 360 (6393), 1096-1101.

[68] Heinz, J., Schulze-Makuch, D., 2020. Tiophenes on Mars: Biotic or Abiotic Origin. Astrobiology 20(4), 552-561.

[69] Cady, S.L., Farmer, J.D., Grotzinger, J.P., Schopf, J.W., Steele, A., 2003. Morphological Biosignatures and the Search for Life on Mars. Astrobiology 3(2), 351-368.

[70] Schwartz, D.E., Mancinelli, R.L., Kaneshiro, E.S. 1992. The use of mineral crystals as bio-markers in the search of life on Mars. Advances in Space Research 12 (4), 117-119.

[71] Masson, A., Masson, G., Rust B.R., 1983. Lacustrine stromatolites and algal laminates in a Pennsylvanian coal-bearing succession near Sydney, Nova Scotia, Canada. Canadian Journal of Earth Sciences 20 (7), 1111-1118, https://doi.org/10.1139/e83-099.

[72] Joseph, R. G., Dass, R. S., Rizzo, V., Cantasano, N., Bianciardi G., 2019. Evidence of Life on Mars? Journal of Astrobiology and Space Science Reviews 1, 40-81.

[73] Williams, A.J., Sumner, D.Y., Alper, C.N., Karunatillake, S., Hofmann, B.A., 2015. Preserved microbial biosignatures in the Brick Flat Gossan, Iron mountain, California. Astrobiology 15 (8), 637-668.

[74] Liu, P., Xiao, S., Yin, C. Zhou, L., Gao, S., Tan, F., 2008. Systematic description and phylogenetic affinity of tubular microfossils from the Ediacaran

Doushantuo Formation at Weng'an, South China. Palaeontology 51, 339-366.

[75] Butterfield, N.J., Knoll, A.H., Swett, K., 1994. Paleobiology of the Neoproterozoic Svanbergfjellet Formation, Spitsbergen. Fossils and Strata 34, 1-84.

[76] Xiao, S., Yuan C., Knoll, A.H., 2000. Eumetazoan fossils in terminal Proterozoic phosphorites? Proceedings of the National Academy of Sciences, USA 97, 13684-13689.

[77] Xue, Y., Tang, T., Yuan C., 1992. Discovery of the oldest skeletal fossils from upper Sinian Doushantuo Formation in Weng'an, Guizhou, and its significance. Acta Palaeontologica Sinica 31, 530-539.

[78] Dong L., Xiao S., Shen B., Zhou C., Li G., Yao J., 2009. Basal Cambrian microfossils from the Yangtze Gorges area (South China) and the Aksu area (Tarim Block, northwestern China). Journal of Paleontology 83, 30-44.

[79] Gray, J., Boucot, A. J., 1989. Is Moyenia a euglenoid? Lethaia 22, 345-456.

[80] Leander, B.S., 2012. Euglenida, euglenids or euglenoids. Version 10. In: The Tree of Life Web Project. http://tolweb.org/Euglenida/97461/2012.11.10

[81] Loeblich, A.R. Jr., Tappan, H., 1978. Some Middle and Late Ordovician Microphytoplankton from Central North America. Journal of Paleontology 52 (6), 1233-1287.

[82] Lyons, T. W., Reinhard, C. T., Planavsky, N. J., 2014. The rise of oxygen in Earth's early ocean and atmosphere. Nature, 506 (7488): 307-315

[83] Ohmoto, H., Runnegar, B., Kump, L. R., Fogel, M.L., Kamber, B., Anbar, A.D., Knauth, P. L., Lowe, D.R., Sumner, D.Y., Watanabe, Y.., 2008. Biosignatures in Ancient Rocks: A Summary of Discussions at a Field Workshop on Biosignatures in Ancient Rocks. Astrobiology, Vol 8, 5, DOI: 10.1089/ast.2008.0257.

[84] Bianciardi, G., Miller J.D., Straat, P.A., Levin, G.V., 2012. Complexity Analysis of the Viking Labeled Release Experiments. Int. J. of Aeronautical & Space Sci. 13(1), 14-26.

[85] Goudge, T. A., Fassett, C. I., Head, J. W., Mustard, J. F., and Aureli, K. L., 2016. Insights into surface runoff on early Mars from paleolake basin morphology and stratigraphy. Geology, 44(6), 419-422.

[86] Zhao, J., Xiao, L., Glotch, 2020. Paleolakes in the Northwest Hellas Region, Mars: Implications for the Regional Geologic History and Paleoclimate. JGR planets, https://doi.org/10.1029/2019JE006196

[87] Grotzinger, J.P., Knoll, E., 1999. Stromatolites in Precambrian Carbonates: Evolutionary Mileposts or Environmental Dipsticks?. Annual Review of Earth and Planetary Sciences, 27:1, 313-358.

[88] Choudhuria, A., Sarkara, S., Altermann, W., Mukhopadhyay, S., Bose, P. K., 2016. Lakshanhatti stromatolite, India: Biogenic or abiogenic? Journal of Palaeogeography, Vol. 5, Issue 3, 292-310.

[89] Rampe, E.B., Morris, R.V., Archer, P.D., Agresti, D.G. Jr., Ming, D.W., 2016. Recognizing sulfate and phosphate complexes chemisorbed onto nanophase weathering products on Mars using in-situ and remote observations. American Mineralogist 101, 678-689.

Chapter 6

# New Insights into the Search for Life on Mars

*Cesare Guaita*

## Abstract

The discovery by the Lander Phoenix (summer 2008) that the Mars polar soil is rich of perchloric acid salts (Na, Mg, Ca perchlorate) strongly could change the interpretation of the Martian experiment of $^{14}CO_2$ release (LR, Labeled release experiment), performed in 70's by both Viking Landers. The LR experiment gave substantially positive results but, at that time, possibility of Martian bacteria was ruled out because the CGMS instruments on board of both Vikings didn't detect any trace of complex organic molecules. But Martian organics exist and were found in fair quantities by Curiosity, landed inside the Gale crater on 2012. So it is likely that Viking CGMS, working at about 500°C, could not see any organic substances (natural or bacterial) because, at that temperature, perchlorates decompose, releasing Oxygen that destroyed organics BEFORE their detection. In any case, the discovery of keragenic compounds by Curiosity, could also be indication of a presence of archea bacteria in the distant past of Mars, when the atmosphere of the Red Planet was wetter and denser than now.

**Keywords:** Mars, organic substances, perchlorate, archea bacteria

## 1. Introduction: from Schiaparelli to Mariner 9

Before the space age, the best Mars observations were performed during the so called oppositions [1]. The distance between the orbits of Earth and Mars varies considerably, largely due to the comparatively large eccentricity of Mars's orbit. Every 780 days on average the Earth overtakes Mars and when, as seen from the Earth, Mars and the Sun are aligned, Mars is said to be in opposition. The opposition distance of Mars from the Earth varies considerably, depending on where Mars and the Earth are in their orbits at opposition. If Mars is near its closest to the Sun (perihelion) the distance is comparatively small, and the opposition is called favourable. Unfavourable oppositions are with Mars near aphelion. The opposition distance varies from 55.7–101 million km, and the corresponding angular diameter of Mars varies from 25.1–13.8 arcsec. Favourable oppositions occur roughly every 15 years.

The first person to draw a map of Mars that displayed terrain features was the Dutch astronomer Christiaan Huygens (**Figure 1**).

On November 28, 1659 he made an illustration of Mars that showed the distinct dark region now known as Syrtis Major Planum, and possibly one of the polar ice caps.

The same year, he succeeded in measuring the rotation period of the planet, giving it as approximately 24 hours.

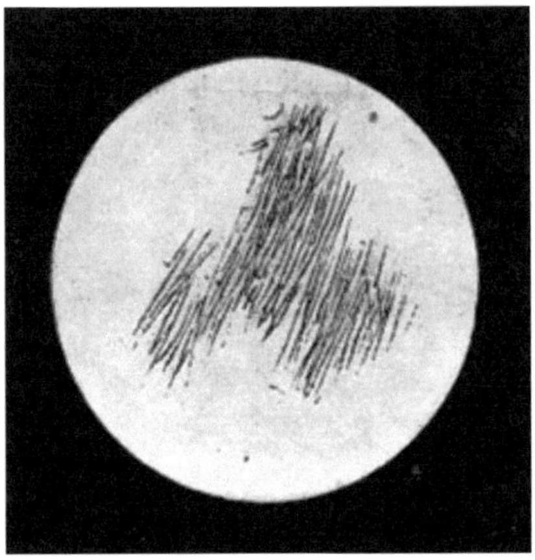

**Figure 1.**
*Syrtis Major, sketched by Christiaan Huygens in 1659. North is at the top.*

In 1666, Cassini detected several distinct dark spots on Mars, and from observing these ascertained that the planet had a rotation on its axis in about 24 hours 40 minutes.

In the oppositions of 1777, 1779, 1781, and 1783, Sir William Herschel determined that the axis of Mars was inclined of about 25° to the plane of its orbit (so having seasons) and measured its polar and equatorial diameters. He showed also that the white spots which formed round the poles of the planet, increased with the approach of winter, and diminished with the approach of summer.

In the oppositions of 1830, 1832, and 1837, Beer and Mädler, observing with a telescope of 4 inches aperture, made a series of drawings from which they were able to construct a chart of the entire globe of Mars. The features which they then drew have been recognised at every succeeding opposition, and some of them can be identified in the rough sketches of Sir William Herschel, and even in those of the year 1666, made by Hooke and Cassini. The surface of Mars therefore possesses permanent features.

In the 1800s, observatories with larger and larger telescopes were built around the world.

A particularly favorable perihelic opposition occurred in 1877 (Mars approcing to within 56 million km on September 5, in Aquarius) [2].

On August 11, 1877 the American astronomer Asaph Hall discovered the two moons of Mars, Phobos and Deimos, using a 660 mm (26 in) telescope at the U.S. Naval Observatory.

In the fall of October 1877, Giovanni Virginio Schiaparelli (1835–1910), scrutinised Mars visually at the Brera Observatory in Milan, where he was director (**Figure 2**). He used a 8′ (21.8 cm, f/14.5) aperture Mertz refractor (magnifying power of 322x), that was installed here in 1865. He named the Martian “seas” and “continents” (dark and light areas) with names from historic and mythological sources [3].

But he is best remembered for about 40 fine lines that he drew crossing the bright red areas, *canali* as he called them. Canali means channels, but it was mistranslated into “canals” implying intelligent life on Mars (**Figure 3**).

Because of the then recent completion of the Suez Canal in 1869 (the engineering wonder of the era), the misinterpretation was taken to mean that large-scale artificial structures had been discovered on Mars. The importance of canals for worldwide commerce at that time without a doubt influenced the popular interest in “canals” on Mars.

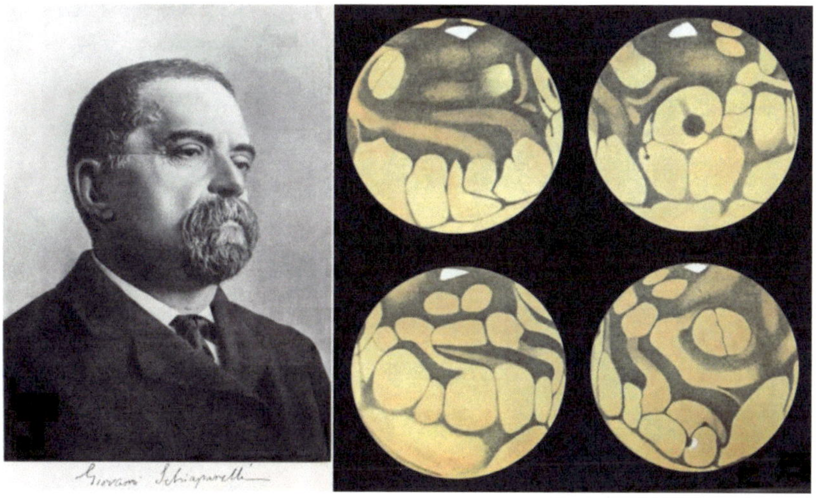

**Figure 2.**
*Original drawings of Mars, made by G.V. Schiaparelli during the 1877 opposition.*

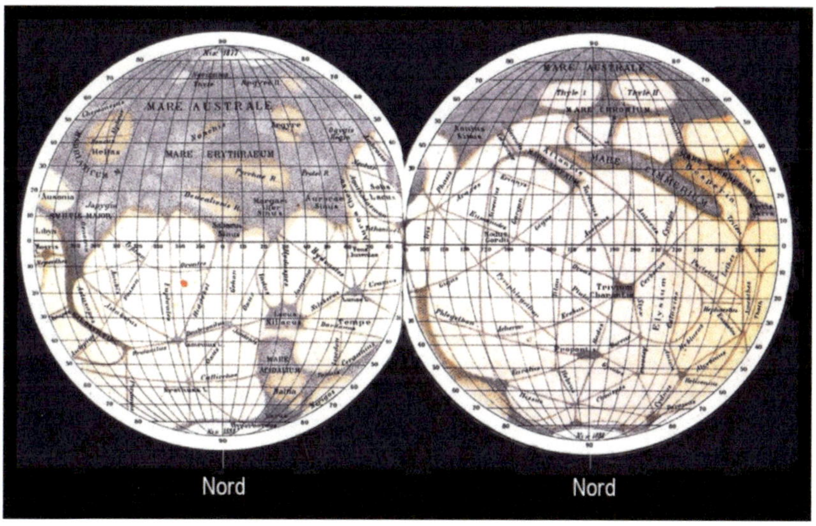


**Figure 3.**
*This drawing of the two hemispheres of Mars was made by the Italian astronomer G.V. Schiaparelli (1835–1910) between the years 1877 and 1888. He named the 'seas' and 'continents' of Mars, and called the straight surface features channels (mistranslated as canals).*

Starting from 1886, Brera was equipped by a larger new instrument, a 16′ (48.7 cm, f/14.3) Mertz-Repsold refractor. Schiaparelli used it, during a couple of following Mars oppositions (1992–1994), to confirm not only the Martian canals but also some duplications of them!

In 1894, Percival Lowel, a member of a rich family from Boston, decided to become a planetary astronomer after reading 'The Planet Mars' a famous book of *Camille Flammarion. He* made his first observations of Mars from a private observatory that he built in Flagstaff, Arizona (Lowell Observatory). He was convinced that the canals were real and ultimately mapped hundreds of them (**Figure 4**).

Lowell believed that the straight lines were artificial canals created by intelligent Martians and were built to carry water from the polar caps to the equatorial regions. In 1895, he published his first book on *Mars* with many illustrations and, over the next two decades, published two more popular books advancing his ideas.

**Figure 4.**
*Mars in 1905, drawn by Percival Lowell. Note the canals. Note also the south polar hood of cloud (at the top).*

Lowell's theories influenced the young English writer H.G. Wells, who in 1898 published *The War of the Worlds*. In this novel, Wells created an invasion of Earth by deadly aliens from Mars and launched a whole new genre of alien science fiction.

On Halloween in 1938, Orson Welles and The Mercury Theater on the Air broadcast a radio version of The War of the Worlds. The story, presented as a series of "live" news bulletins, panicked thousands of listeners who believed that America was being attacked by hostile Martians.

In the 1953, the story of The War of the Worlds of H. G. Wells was adapted in a famous American science fiction film from Paramount Pictures, produced by George Pal and directed by Byron Haskin. Earth was suddenly and unexpectedly invaded by Martians. Many of the major world capitals were destroyed by the Martians, being Martians impervious to all humanity's weapons (enclosed an atomic bomb!) . But, after all that men could do had failed, the Martians were destroyed by terrestrial bacteria to which only mankind have long since become immune.

The real origin of the Mars canals was revealed by another Italian astronomer, Vincenzo Cerulli (1859–1927), founder in 1890 of the Astronomical Observatory in Teramo (in the Abruzzo region), equipped by a Cook Refractor of 40 cm. Starting from 1897, he gave a convincing explanation, still generally accepted. He suggested that the lines were a sort of optical illusion, created by the human brain that "needs" to interpret even vague and indistinct images with familiar shapes. Therefore, poor quality images, such as those that low quality telescopes would provide, would be interpreted as structured shapes, for example connecting individual roughly aligned "dots" into straight lines. This has been demonstrated by many laboratory and field experiments.

During the great opposition of 1909, on the night of September 2020, thanks to exceptional seeing conditions, Eugène M. Antoniadi (1870–1944), one of the most skilled observer of his time, using the new big 83 cm reflector of Meudon observatory, gives some exquisite drawings of Mars, in which all "canals" having some feedback in the past vanished (**Figure 5**).

On 15 July 1965 the NASA spacecraft Mariner 4 flew past Mars, at a minimum distance of only 9800 km. This first Mars flyby gave a major negative surprise: all 22 of the images sent to Earth showed a cratered landscape more akin to the Moon (**Figure 6**). The heavily cratered (and hence ancient) surfaces indicate lack of

**Figure 5.**
*In 1909, E.M. Antoniadi got the chance to use the great 33" refracting telescope at Meudon, on the outskirts of Paris, and on the first night, Sep. 20, he saw Mars so clearly that he could not believe his eyes. It was covered with detail, but not a bit of it was geometric–there was not a canal in sight.*

**Figure 6.**
*Images sent to Earth by Mariner 4 and Marine 6 during their flyby of Mars in July1965 and July 1969.*

geological activity and lack of extensive weathering by water, which would have erased these craters in a fraction of the age of Mars. Moreover, when Mariner 4 passed beyond Mars as viewed from the Earth, the changes induced by the atmosphere enabled the surface pressure of the atmosphere to be determined: the value was a mere 6 millibars, ten time less than previously believed. A pressure of 6 millibars is close to the triple pressure of water, below which water cannot exist in a stable liquid phase at any temperature. Because all life on Earth requires liquid water, the prospect of finding life on Mars faded a lot.

The next two successful missions to Mars, were the flyby (at a distance of about 3400 km) of Mariner 6 on July 31, 1969 and the flyby of Mariner 7 a few days later on August 5, 1969, flying by over the equator and south polar regions and analyzing the Martian atmosphere and surface with remote sensors, as well as recording and relaying hundreds of pictures. The two spacecraft returned a combined total of 143 approach pictures of the planet and 55 close-up pictures (**Figure 7**).

Again the small amount of martian surface investigated was found covered by impact craters. More, the temperature of the south polar cap was measured and found to correspond to the solid–gas phase boundary of $CO_2$ at a pressure of a few millibars. This provided strong evidence that a polar cap of $CO_2$ was roughly in equilibrium with the $CO_2$ atmosphere.

Subsequent studies have confirmed that the seasonal cap at both poles is indeed predominantly $CO_2$ snow and frost, but that this overlies a permanent cap mainly composed of dusty water ice at the North Pole, and dusty $CO_2$ ice at the colder South Pole, perhaps underlain by dusty water ice.

The results of Mariner 4,6,7 were very disappointing. But it was soon realized that some fast flybys over no more than 20% of the martian surface, could not give a satisfactory knowledge of a complex planet as Mars. It would have been necessary to map the *whole* planet and make continous obsevations for a long period of time, objectives that only an orbital mission could carry out. The Mariner 9 mission was born.

On November 14, 1971 the spacecraft Mariner 9 was placed, for the first time in the history, in orbit around Mars. But, unbelievably, Mariner 9 arrived when Mars was obscured by the largest dust storm ever observed. The surface was totally obscured for a couple of months and the imaging program did not get underway until mid-January 1972. The spacecraft was turned off on October 27, 1972, after 349 days in orbit, and 7,329 images transmitted, covering 85% of Mars' surface. With two astonishing discoveries (**Figure 8**).

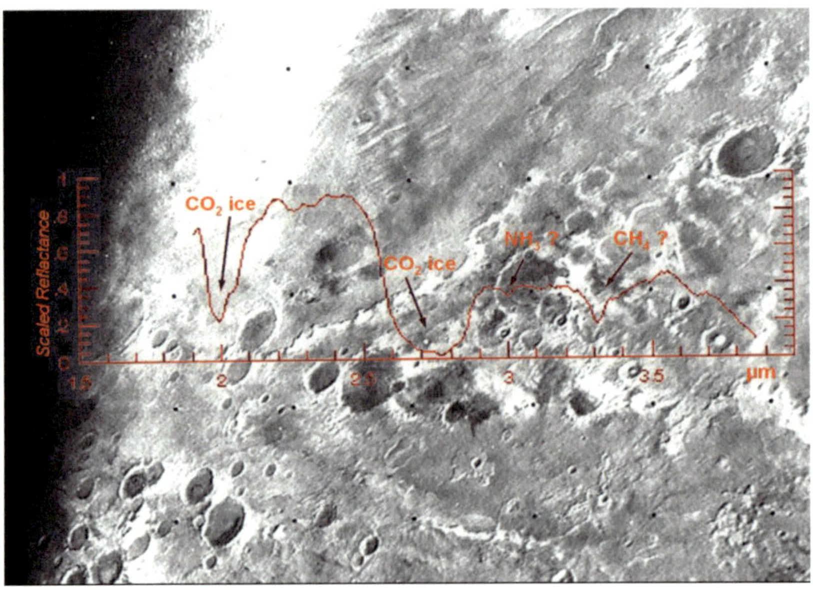


**Figure 7.**
*South polar cap of Mars as seen by Mariner 7 in August 1969. The IRS spectrometer on board of Mariner 7 demonstrated the presence of $CO_2$ ice and, possibly, also trace of $NH_3$ and $CH_4$ [4].*

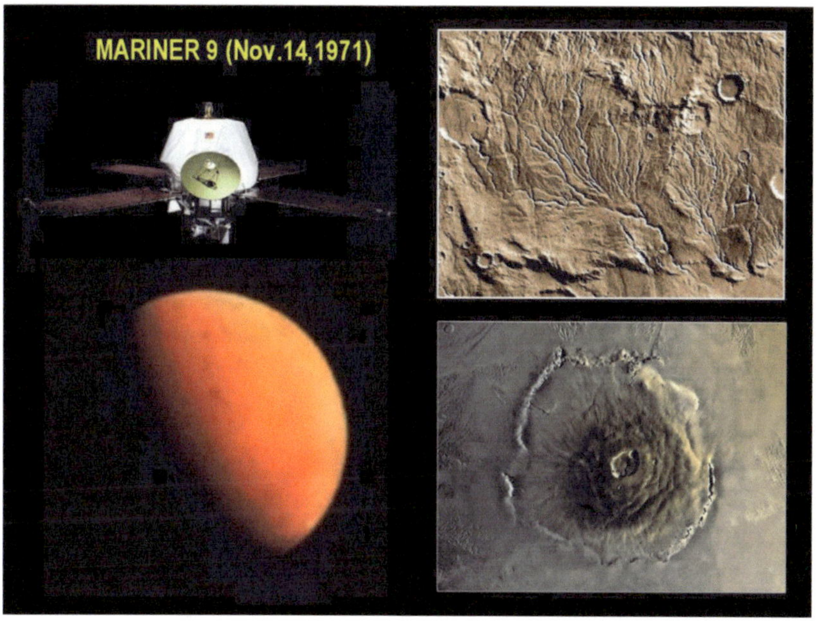


**Figure 8.**
*The largest dust storm ever observed obscured all the surface of Mars in November 1971, when Mariner 9 entered orbit for the first time. When the storm subsides, Mariner 9 made two main discoveries: big volcanoes and hundreds of extinct riverbeds.*

The first main discovery was the existence of several huge 'young' shield volcanoes, so high that their peaks emerged also from the dust storm. Most of these volcanoes were located in the Tharsis Regio, a vast plateau 4.000 m high, centered near the Martian equator, that formed about 2 billion years ago, giving rise to an enormous canyon system, named Mariner Valley, after Mariner 9 in honor of its achievements. The second main discovery was that of hundreds of extinct riverbeds, that seem to have been carved by the flow of liquid water early in Martian history: an indication that Mars was much warmer and wetter in the past. Certainly most exciting discoveries as far as life on Mars is concerned, that convinced NASA to plan a mission to directly search for life on Mars. The Viking program was born.

## 2. The intringuing results of the Vikings program

The two Viking Landers in the '70 years made the first direct search of traces of present or past biological life on Mars [5] (**Figures 9** and **10**).

The results were so contentious that, after more than 45 years, no unambiguous interpretation was found.

The GCMS instruments (Gas Chromatograph-Mass Spectrometer) on board both Viking Landers [6, 7] were tasked with detecting organic compounds. GC–MS heated many samples of martian soil up to 500°C, but did not detect any trace of complex organic molecules, even if detected an amount of 0,1–1% of $H_2O$ and 50–500 ppm $CO_2$ respectively (**Figure 11**) and the enigmatic release of about 15 ppb of $CH_3Cl$ (chloro-methane) and up to 20–30 ppb of $CH_2Cl_2$ (methylene chloride) (**Figure 12**).

At that time, the two light chloro-derivatives, being released together with some trace of a solvent of sure terrestrial origin such as Freon-E, were considered as a terrestrial contamination, ruling out the occurrence of any form of martian life [8]. In the meantime the $H_2O$ and $CO_2$ release upon heating were explained as thermal decomposition of hydrous silicates and carbonates respectively.

**Figure 9.**
*Viking 1 on the surface of Cryse Planitia. The white arm in the center is a meteorological sensor; the arm that took samples is visible to the right.*

**Figure 10.**
*May 18, 1979: water frost on Mars rocks and soil near the Viking 2 lander.*

The measured ratio 37Cl/35/Cl = 0,319 similar to that one of the terrestrial chloride supported to this interpretation. However, Z.D. Sharp [8] found that the ratio 37Cl/35/Cl is quite constant all over the Solar System: actually its value is the same on the Earth, on the chloro-salts enclosed inside the carbonaceous chondrites and also on some Martian meteorites.

Moreover, a suitable inquiry proved that $CH_3Cl$ and/or $CH_2Cl_2$ never were used during the Viking assembly, so any trace on board was impossible [9].

We could re-discuss the conclusion taken from the Viking GC–MS results on the basis of a couple of reasons.

The first reason is that the Martian soil in any case should be enriched by the organic molecules (that is carbon-containing chemicals) that could be taken by comets and carbonaceous chondrites. The recent discovery of simple and polymeric organic substances inside many Martian meteorites could be an evidence in this regard [10].

Carbon-containing chemicals such as those that make up the stuff of life on Earth, had been found in rocks that were blasted off Mars millions of years ago by large asteroid impacts and fell to Earth as meteorites (at present about 300 Martian meteorites are known). But no one could be sure the organics in Martian meteorites

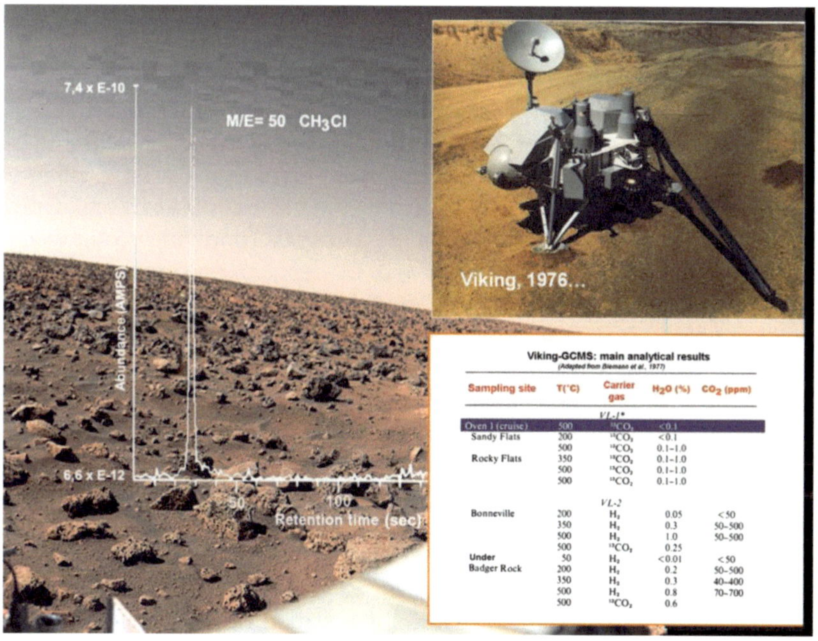


**Figure 11.**
*The main analytical results of the CGMS (GasChromatograph-Mass Spectrometer) on board of the Viking landers.*

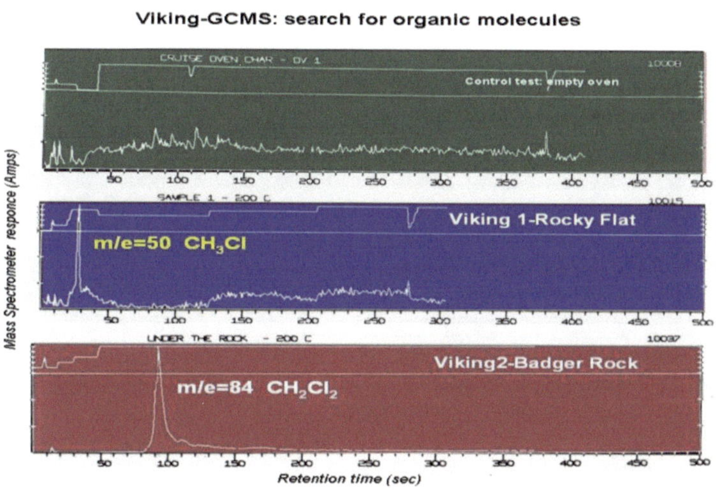


**Figure 12.**
*The GC on board of the Viking Landers detected only enigmatic presence of chloro-methane and methylene chloride, at that time deemed terrestrial pollutants.*

weren't just earthly contamination. So microbiologists of the Carnegie Institution for Science's Geophysical Laboratory in Washington, D.C., looked in the most protected parts of Martian meteorites: microscopic mineral grains that had been securely locked in larger crystals for up to 4 billion years. Using a number of analytical techniques, they probed for organic matter through the encasing minerals. Organic chemicals were found in 10 of 11 once-molten meteorites examined, at an abundance of about 20 parts per million. Raman spectroscopy showed that they include large, complex carbon compounds rich in benzene-like rings of carbon atoms, included polycyclic aromatic hydrocarbons (PAHs), typical of organic-rich meteorites such as carbonaceous chondrites. Given the way the organic matter was sealed in the rock, "it is carbon from Mars," not terrestrial contamination.

A special case remains that of ALH 84001, (**Figure 13**) a meteorite found in 1984 in the antartic region of Allan Hills. Ejection from Mars seems to have taken place about 16 million years ago. Arrival on Earth was about 13 000 years ago. Cracks in the rock appear to have filled with carbonate materials (implying groundwater was present) between 4 and 3.6 billion-years-ago. Evidence of polycyclic aromatic hydrocarbons (PAHs) have been identified with the levels increasing away from the surface. In the crack were also found deposits of iron as magnetite, that are claimed to be typical of bio-depositation on Earth [11].

In some SEM (Scanning Electronic Microscope) pictures taken inside the carbonate material, small ovoid and tubular structures were found [12], that might be interpreted as nanobacteria fossils (**Figure 14**), but also as sample preparation artifacts, being, at that time, unknown earthy life forms so small. The controversy has never ceased even if, some years after, living colonies of nano-organisms were found on Triassic and Jurassic sandstones from petroleum exploration wells offshore western Australia [13]. These living forms were up to 10 times smaller in diameter (10 nm) compared to 'normal' cellular structures.

The other reason for which the negative response of the two GCMS onboard the Viking Landers has to be rediscussed is linked to the fact that the Viking Labeled Release (LR) experiment, made at ambient temperature (10–15°C), gave a biological result substantially positive (Levin, 1976). LR it has been the only experiment with a clearly positive response, whereas the other two 'biological' experiments, i.e. the Gas Exchange (GEX) experiment [14] and the Pyrolytic Release (PR) experiment (Carbon Assimilation Experiment) [15] gave dubious results, suggesting a lack of biological reactions.

**Figure 13.**
*Inclusions of carbonates inside the Martian meteorite 84001, found in Antartica in 1984.*

**Figure 14.**
*The tube-like forms on this highly magnified SEM image of ALH84001 could be fossils of simple Martian organisms that lived over 3.6 billion years or, simply artifacts of sample preparation.*

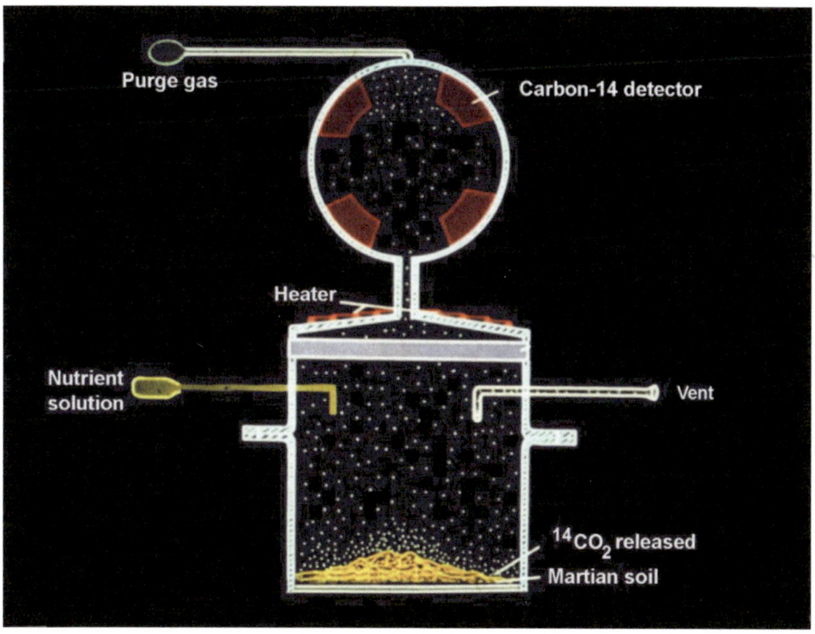


**Figure 15.**
*A scheme of the LR experiment (Labeled Release) on board of Viking Landers.*

The LR experiment (**Figure 15**) was based on the well-known concept that all terrestrial microorganisms metabolize the organic substances releasing $CO_2$.

In the Viking LR experiment, the Landers collected samples of Martian soil by means of their robotic arm, injected them with a drop of dilute nutrient solution containing alanine, formic acid, glycine, glycolic acid and lactic acid, and then monitored the air above the soil for signs of metabolic byproducts. Since the nutrients were tagged with radioactive carbon-14, if microorganisms in the soil metabolized the nutrients, they would be expected to produce radioactive byproducts, such as radioactive carbon dioxide ($^{14}CO_2$). $^{14}CO_2$ was indeed released when an aqueous solution of 14C labelled amminoacids was added, but a much lower amount (i.e by more than one order of magnitude) was released in the case of terrain samples sterilised at 160°C (**Figure 16**). To rule out the possibility that the strong ultraviolet radiation on Mars might be causing the positive results, the Landers collected also soil buried underneath a rock, which again tested positive. The control tests also worked, with the 160°C sterilization control yielding negative results [16].

In 2002 a possible circadian fashion (i.e. having the same periodicity of the Martian day) of $^{14}CO_2$ release was found, which may be a typical biological signature [17]. A complex statistical analysis [18] reached the same conclusion.

In any case, it is important underline that the same Levin [19] observed that the release of radioactive $CO_2$ could be due also to nonbiological reactants, a real possibility discussed also by Klein [20]. On this subject some lab tests were performed, assuming that the Fe superoxides are built up in the Martian soil by the strong UV radiation. This Fe superoxide could decompose (with the release of $^{14}CO_2$) the carbon molecules of the LR nutrient solution directly [21] or through the formation of $H_2O_2$ [22]. In the presence of water, the superoxide ion reacts to produce Oxygen, perhydroxyl radical, and hydroxyl radical [23]: $2O_2^- + H_2O \rightarrow O_2 + HO_2^- + OH^-$.

This release of oxygen could decompose carbon molecules in the LR experiment, but could also explain the results of the Gas Exchange (GEX) esperiment. In GEX a water solution of many amino-acids and salts, was injected into a sample of martian soil, measuring by GC (Gas Cromatography) any gas emission such as the release of

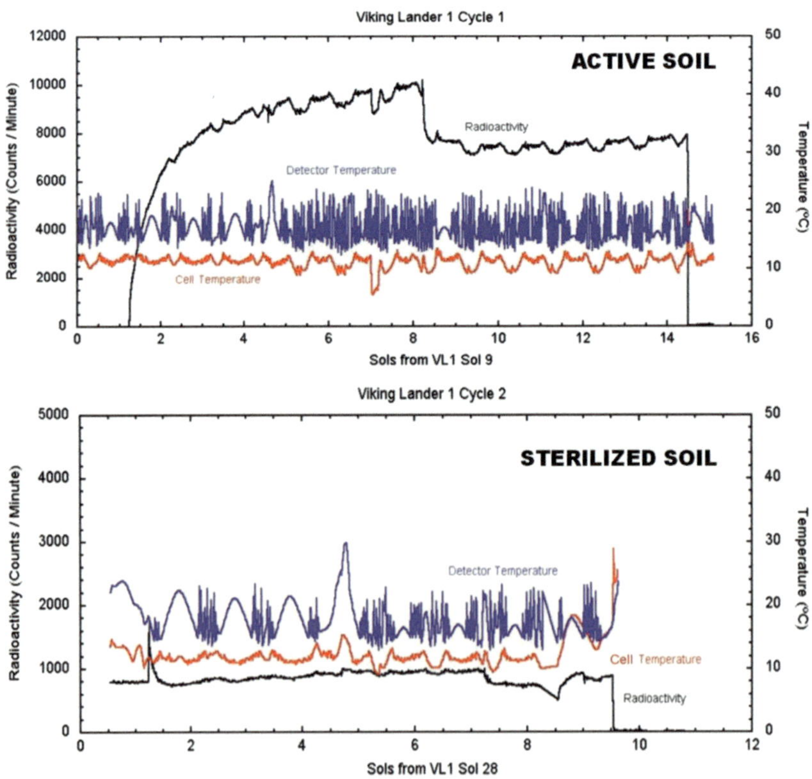


**Figure 16.**
*The Viking 1 result of Labeled Release experiment on active and sterilized Martian soil (see the text for details).*

**Figure 17.**
*June 25, 2008: a trench, called 'Dodo-Goldilocks,' showing lumps of water ice in this picture taken by the Phoenix Lander. A big surprise was the discovery of perchlorates by the MECA instrument.*

$H_2$, $N_2$, $0_2$, $CH_4$, Kr, and $CO_2$. Analyses showed that a 1 $cm^3$ of Martian soil sample produced up to 700 nM of Oxygen after 50 hours, an amount far superior to any terrestrial test, and so not believed to be of biological origin.

A possible explanation of the intringuing results of the Viking analyses was found by the Wet Chemistry Laboratory on the Phoenix Mars Lander (**Figure 17**) that, in the summer of 2008, discovered up to 0,6% of Magnesium perchlorate - Mg $(ClO_4)_2$ in the North polar sands of Mars [24]. This salt is inert at low temperature, but at high temperature became a strong oxidant able to decompose all carbon compounds.

So, if we suppose that also the soil sampled by Viking were rich of perchlorates, the GCMS analysis, being performed at 500°C, possibly could results in a demolition of all organic molecules (biological or not) during the same analytical process. The assumption of the presence of perchlorates at the Viking landing sites might seem a little hasty, because perchlorate may form preferably at high latitudes [24], whereas the Viking 1 landing site was at equatorial latitudes, and the Viking 2 one at intermediate latitudes. However, we could not exclude this possibility, specially after the discovery of perchlorates inside the Gale Crater (Lat = 5,24° S), landing site of Curiosity [25] and the discovery of perchlorates also in some martian meteorites [26].

In the case of analyses performed at low temperature, perchlorates are totally inert and so a positive response, as observed by the Viking LR experiment, may really suggest the presence of organic substances.

## 3. Viking and perchlorates

A further support to a biological interpretation of the Viking LR experiment was given by R. Navarro-Gonzales [27]. In summary, a sample of a Mars-like soil of the driest core of the Atacama desert in Northern Chile (The Yungay Area), containing very low organic concentration (32 ppm), was subjected to a thermal volatilization process. The released gases and volatiles have been then measured by a GCMS similar to the Viking ones. At 500°C a clear emission of organic substances such as benzene, toluene, formic acid was observed. But when the same soil was heated at 500°C after the addition of 1% of Mg perchlorate, the organic substances mentioned were no longer observed, whereas release of $CO_2$ and $H_2O$ and, amazingly, also of $CH_3Cl$ and $CH_2Cl_2$ was measured (**Figure 18**).

According to Navarro Gonzales [27] the release $CH_3Cl$ and $CH_2Cl_2$ was ascribed to a reaction between perchlorate and organics. According to experiments on simulated Martian soil [28], the thermal action of perchlorate in the Vikings GCMS results should have substantiate by the detection of some chlorinated aromatics (such as chlorobenzene and chlorotoluene). Well, an accurate re-examination of the original, microfilm preserved, Viking GCMS data sets [29] found evidence for the presence of chlorobenzene in Viking Lander 2 (VL-2) data, at levels corresponding 0.08–1.0 ppb, in two sample heated to 350°C and 500°C. A surprising discovery that is also a demonstration of the presence of perchlorate in the Viking martian soil.

Unfortunately, the two Vikings were not able to search for perchlorates. But it is possible 'to read' the potential presence of these salts in some meaningful clues.

For example, the RXFS (X-Ray Fluorescence Spectrometer) on board the Viking lander was suitable to search for Cl (Chlorine) in martian soil, finding similar values: Viking 1 found 0,8% of Cl on the landing site of Cryse (22,7°N, 48,2° W) and Viking 2 found about 0,4% of Cl on the landing site of Utopia (48,3°N, 226°W) [30, 31]. Furthermore Pathfinder (1997) found up to 1% of Cl on Ares Valley (19,3°N, 33,6°W) [32], Spirit found about 0,5% of Cl inside the Gusev crater (14,6°N, 175,5°E) [33] and Opportunity found up to 1% of Cl on Meridiani Planum (1,9°S, 354,5°E) [34]. A more general investigation was made by the orbital spacecraft Odissey 2001, between June 2002 and April 2005. Its Gamma Ray Spectrometer (GRS), measured the equatorial and mid-latitude distribution of Cl at the near-surface of Mars, finding a not homogeneously concentration, with a mean value of 0.49 wt% Cl and variation up to a factor of 4 [35]. The kind of compound containing Cl should be investigated.

After the unexpected discovery of up to about 1% of Mg-perchlorate on the Martian polar soil (68.3°N, 127.0°W) performed by Phoenix Lander [24], with only

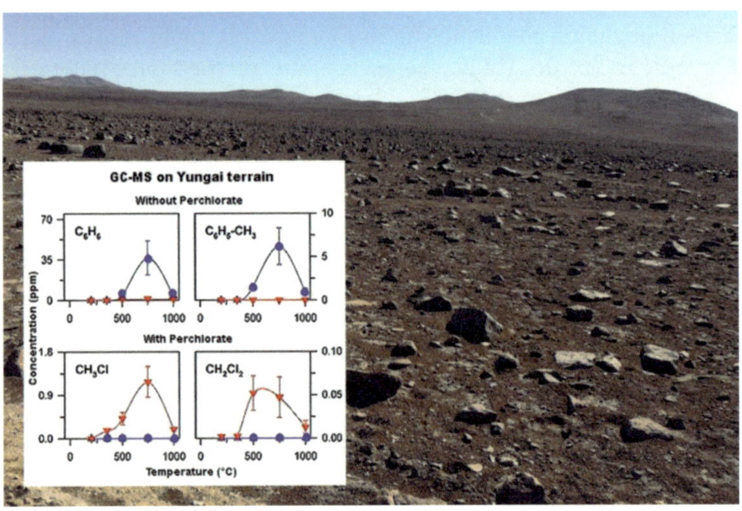


**Figure 18.**
*The dry soil of the Yungay Area (Northern Atacama), if heated up to 500°C with the external addition of perchlorates, shows the same behaviour of the Martian soil: disappearance of organic signals and release of light chloro-hydrocarbons.*

traces (0,02%) of other salt containing Cl, the quite abundant amount of Cl found by Vikings appears as a strong indication of the presence of perchlorates. The mechanism of Martian perchlorate production is still being debated. It has been suggested that production pathways for perchlorate on Mars are similar to Earth, primarily photochemically in the upper atmosphere via oxidation of chlorine by ozone [36]. But because of the low amount of Ozone in the Martian atmosphere, mechanisms involving surface components are probable [37]. For example, perchlorates may form from the radiolysis of surface component caused by galactic cosmic rays, causing a sublimation of chlorine oxide in atmosphere, where final oxidation to perchloric acid is performed by some sources of active Oxigen (i.e. $O_3$ and/or $CO_2$ photolysis) [38]. And in the presence of a suitable catalist such as $TiO_2$ the strong Martian UV illumination could oxidize chloride ions to perchlorate also in aqueous solutions [39].

The permanence of perchlorates (very soluble in water) on the martian soil is made possible by the strong ambient dryness: Mars lacks rains able to dissolve perchlorates for millions of years.

The logical interpretation of the Navarro Gonzales [27] results on the Atacama soil starts from the well known decomposition of $Mg(ClO_4)_2$ at temperature > 400°C [40], with release of $O_2$ and Cl:

$$2\,Mg\,(ClO_4)_2 \rightarrow MgO + MgCl_2 + 15/2\,O_2 + 2Cl$$

$O_2$ and Cl react with organics compounds, releasing, on one side, $H_2O$ and $CO_2$, and, on the other side, the chlorine compounds observed by the Viking GCMS.

The results of Phoenix and Atacama analyses, suggested to reconsider methods for searching carbon molecules on Mars, taking in account the significant risk arising from the thermal methods.

## 4. Curiosity and SAM results

The first chance for to this new approach occurred with the Curiosity mission (NASA, Mars Science Laboratory Press Kit, 2012), a rover of 900 kg that landed successfully on August 6, 2012 inside the Martian Gale crater (5.4°S 137.8°E) at a lower latitude than Viking (Cryse at 22,7°N and Utopia at 48,3°N), an ancient lake, with a layered mountain 5,000 m high in the center (the Mount Sharp). The task of Curiosity was to reach the mountain and to climb on it, in order to disclose the geological past of Mars, starting from the farter past (lower stratification) (**Figure 19**). The most interesting soils were found right at the base of Mount Sharp, where Curiosity encountered a dangerous expanse of dark sand (Bagnold Dunes), a long ridge rich of hematite (Vera Rubin Ridge), a clay-bearing unit (Glen Torridon), followed by Sulfur-rich uneven ground.

On January 1, 2018 (sol 1992) near the southern edge of the Vera Ruin Ridge, the Mars Hand Lens Imager (MAHLI) camera on Curiosity movable arm, pointed out a cluster of millimetric dark, stick-shaped features whose origin is uncertain. One possibility is that they are erosion-resistant bits of dark material from mineral veins cutting through rocks in this area (**Figures 20** and **21**).

But the morphological analogy with terrestrial fossil traces of life-substrate interactions is impressive [41]. Some studies even highlight occurrence, on Martian sediments, of widespread structures like the famous microspherules discovered by the Rovers Spirit and Opportunity, often organized into some higher-order settings (**Figure 22**). Such structures also occur on terrestrial stromatolites in a great variety

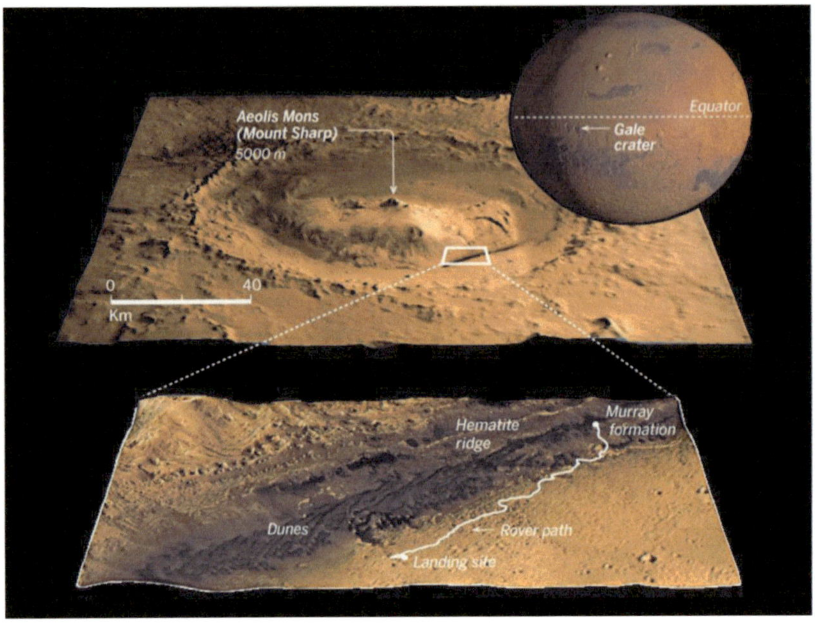


**Figure 19.**
*The path of Curiosity inside the Gale crater, to reach the base of Mount Sharp.*

**Figure 20.**
*December 13, 2017 (sol 1903): Curiosity near the Vera Rubin Ridge, a formation very rich of hematite.*

of microscopic structures, such as voids, gas domes and layer deformations of microbial mats [42].

SAM (Sample Analysis at Mars) is a suite of instruments aimed at analyzing of soil samples on board of Curiosity (**Figure 23**).

It includes an improved and more sensitive (up to 100 times) version of the Viking GC MS [43] and a laser infrared spectrometer (TLS) [44] able to analyse any gaseous substance from both Martian atmosphere and GC–MS, with a sensitivity of 1 ppb (part per billion). The SAM works by accepting drilled or scooped Martian grit into a tiny cup made of quartz, that can be cooked in an oven up to 1100°C. Tiny puffs of helium gas move the gases from the sample cup into a MS (Mass Spectrometer), that sift through the resulting fumes for molecular signatures, directly (EGA, Evolved Gas Analysis) or after a previous separation inside one of 6 column GC (Gas Chromatographic) (**Figure 24**). There are 74 cups in a carousel: 59

**Figure 21.**
*January 1, 2018 (sol 1992): these enigmatic dark, stick-shaped features taken from the MAHALI camera on board of Curiosity look alike terrestrial fossils of the Ordovician period.*

**Figure 22.**
*Examples of mesostructural parallels with occurrence of spherical bodies having similar shape and structure. On Mars, 'blueberries' could assume polycentric polispherules or concentric structures. Such parallels occur also for living colony of cyanobacteria (frame II), as polispherule and concentric structures and for stromatolites (the knot structure in the frame III.*

are quartz tubes slated for "dry" chemistry, 9 are solvent cups, sealed with foil, contain solvents for 'wet' chemistry to tease out organic molecules, like amino acids and degraded fatty acids, that would otherwise resist vaporization; the last 6 are calibration cups (**Figure 25**).

Among other capacities, the TLS [44], detecting the IR absorption band of $CH_4$ at 3,27 micron, was aimed at confirming the existence and the seasonal cycle of methane discovered by terrestrial telescopes [45] and possibly confirmed by PFS

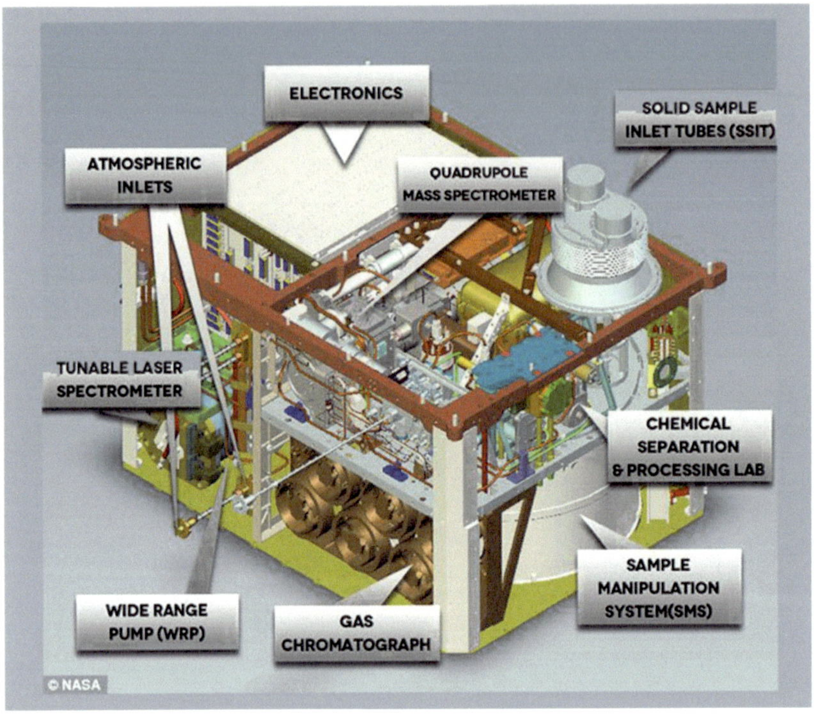


**Figure 23.**
*The suite of instruments of SAM (Sample Analysis at Mars) laboratory on board of Curiosity.*

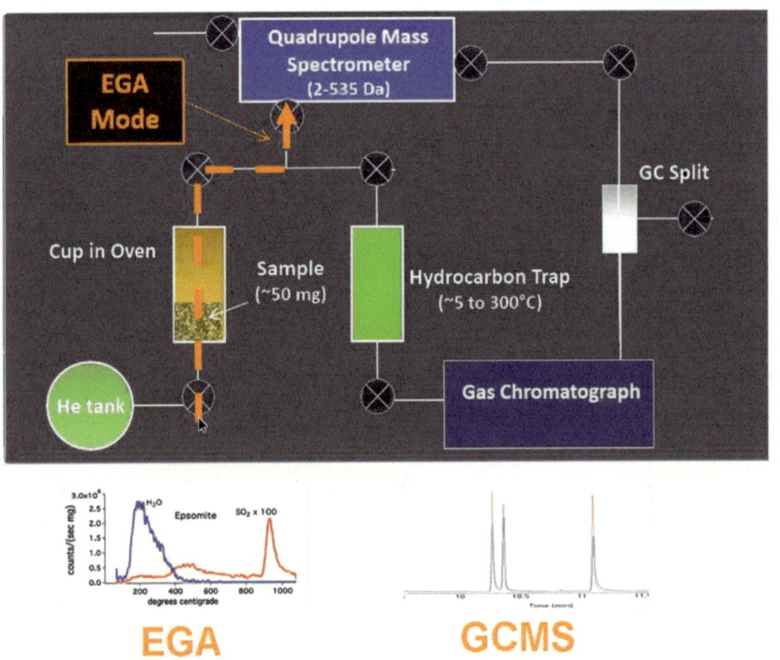


**Figure 24.**
*Inside the SAM lab., gases released by a sample of Martian soil heated up to 1000°C can be sent to a Mass Spectrometer (MS) directly (EGA-MS, Evolved Gas Analysis) or passing before though Gas-chromatographic column (GCMS).*

DOI: http://dx.doi.org/10.5772/intechopen.97176

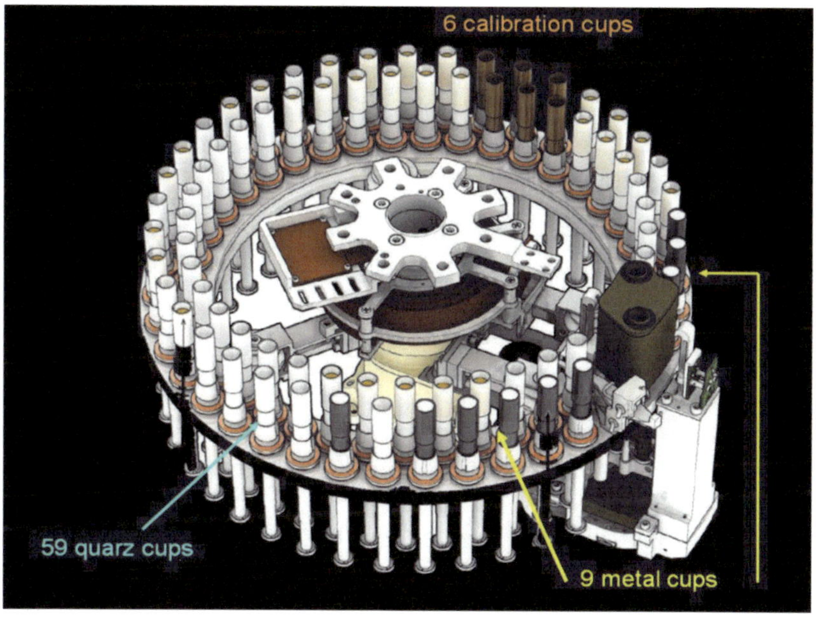


**Figure 25.**
*The Carousel of SAM, called the SMS (Sample Manipulation System). It contains 74 cups, dedicated to receiving solid samples collected by the Curiosity rover. The 74 sample cups are separated into three categories: 59 solid sample quartz cups, 9 foil topped metal cups for wet chemistry experiments (7 with MTBSTFA, 2 with TMAH), and 6 foil topped cups of reference samples.*

spectromenter on board of Mars Express [46]. The task of TLS was to continue the search for methane in order to establish its source (geological or biological).

Really, TLS detected methane many times over the course of the mission, though with a very strange behaviour. Background levels of the gas seem to rise and fall seasonally (0,24–0,65 ppbv, parts per billion units by volume) [47] (**Figure 26**).

The highest methane levels do appear just after the warmest time of the year, suggesting that heat spreading downward allows more of the gas to be released.

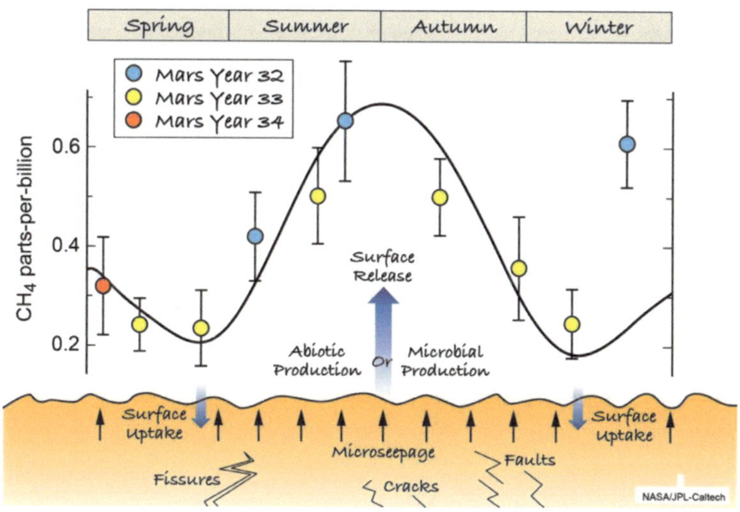


**Figure 26.**
*Seasonal variations of Martian methane, detected by the SAM-TLS spectrometer, during 55 months (about three Mars years from March 17 2014 to March 21, 2017.Potential methane sources include methanogenesis by microbes, ultraviolet degradation of organics, or water-rock chemistry. The methane could be later destroyed by atmospheric photochemistry or surface reactions, as examples. Seasons refer to the northern hemisphere.*

Finding methane in Mars's atmosphere is intriguing because chemical reactions should destroy the gas after about 300 years. So its presence today suggests that something on the planet is still sending the gas into the atmosphere. The source could be geological, such as reactions between certain types of rock and water (basaltic serpentinization) or could be linked to ancient methane trapped in clathrate hydrates; more intriguingly, the warmest season could 'awaken' buried microbes or other forms of life, taking in account that most of the methane in Earth's atmosphere comes from living processes. But a recent statistical analysis [48] casts doubt on the hypothesis of "seasonal variability" in Mars' surface methane, finding that it is unsupported by the Curiosity TLS data. This is because the data are too sparse over too limited timespan, to favor a seasonally cyclic explanation of the data over alternative hypotheses of stochastic variation or variation with other periods.

TLS detected also episodically increases ('spike') of Martian methane [49]: for example on June 16, 2013 and on early January 2014 readings averaged ten time the background level (6–8 ppbv). The largest concentration of methane detected in situ by the Curiosity reached a spike to 21 ppbv, on June 20, 2019, dropping quickly over a few days (**Figure 27**).

The PFS spectrometer of Mars Express found a possible geological origin of this n usual pattern [50]. Indeed, in a re-examination of archive data, PFS, on June 16, 2013, observed an elevated spot level (15.5 ± 2.5 ppbv) of methane, from a nearby area called Medusae Fossae, located about 500 km east of Gale crater. The Mars Express observation was made 20 hours before the methane spike of 5.78 ± 2.27 ppbv reported by TLS-SAM. Being Medusae Fossae a fractured and likely volcanic in origin, it is possible that a therein geological emission of methane has been carried by the prevailing winds towards the Gale crater (**Figure 28**).

Highly sensitive measurements of the atmosphere of Mars performed by the ESA-Roscosmos ExoMars TGO (Trace Gas Orbiter) from April to August 2018 made the problem of Martian methane even more enigmatic [51]. No trace of methane was indeed found by two instrument suites onboard TGO designed to perform such measurements: ACS (the Atmospheric Chemistry Suite) and NOMAD (Nadir and Occultation for Mars Discovery) that cover the 3.3 μm spectral range,

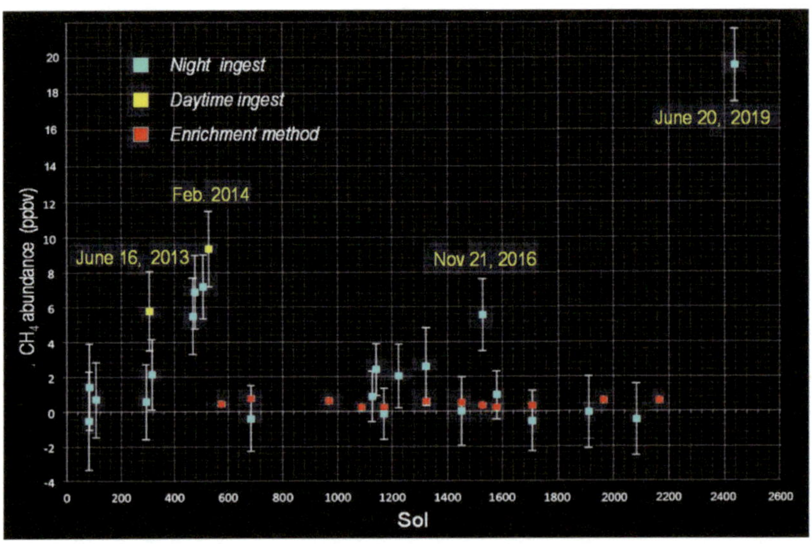


**Figure 27.**
*TLS-SAM methane measurements at Gale crater over an ~4.5 Earth years (~56 months) period (from 26 October 2012 to 27 May 2017), taken during the rover's journey of 16.5 km over highly varied terrain.*

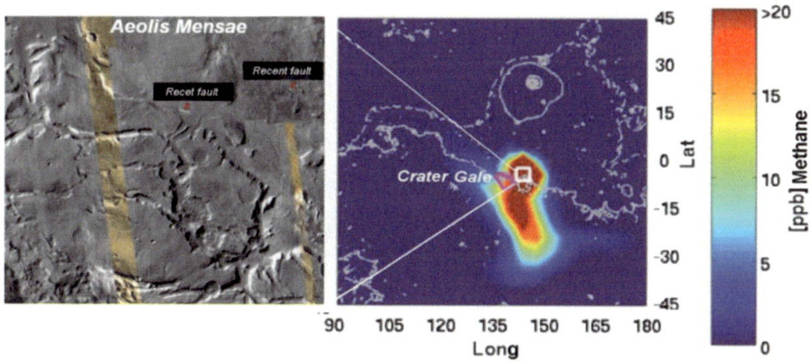


**Figure 28.**
*On June 13, 2013, the PFS spectrometer on board of Mars Express observed a methane emission over the fractured terrain of Medusae Fossae, some hours before a spike of methane detected inside the Gale crater by the TLS-SAM instrument. Geological methane carried towards the Gale crater by the prevailing winds?*

which includes the strongest fundamental absorption bands for hydrocarbons such as $CH_4$, in particular the $\nu 3$ asymmetric stretching band on which all the previous detections were made.

Until the end of 2020 the SAM-GCMS made more than twenty complete analyses on Gale crater soil (**Figure 29**).

Inside the Gale crater Curiosity discovered for the first time, Martian organic molecules, just after a few attempts (Rocknest, John Klein, Cumberland at Yellowknife Bay, not far from the landing site) (**Figure 30**).

Yellowknife Bay mudstone is thought to contain sediments transported by fluvial and deltaic processes from the crater rim area to the north.

Between sol 56 and 100 (October 2 to November 16, 2012) Curiosity reached the sandy terrain of Rocknest, located about 550 meters away the landing site. The APXS instrument (Alpha Particle-X rays spectrometer) [52] detected on Rocknest a little amount of S and Cl [53]. The sandy texture of the soil was suitable to be easily transferred inside the SAM. Under the heating of the sample up to 800°C, many kinds of gaseous substances were released [54]. The release of molecular Oxigen ($O_2$) at 300–400°C (**Figure 31**) was very important: together with the presence of

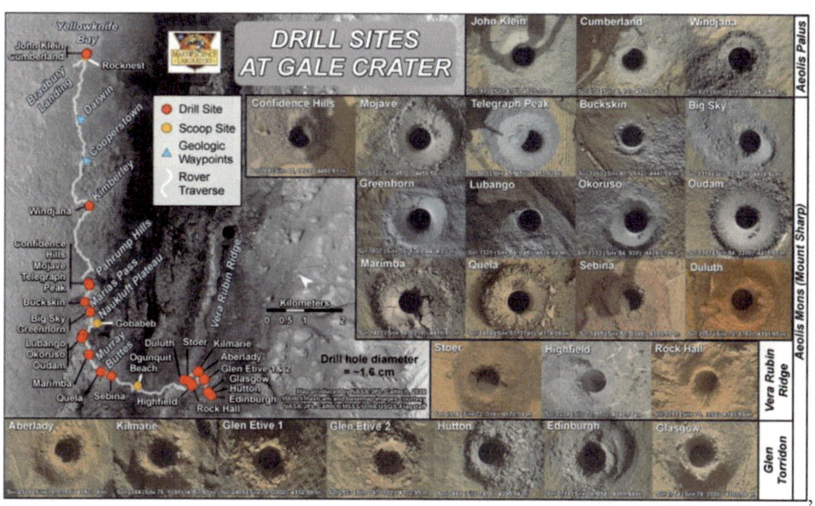


**Figure 29.**
*A summary of all drill sites made by Curiosity at Gale Crater up the end of 2020.*

**Figure 30.**
*December 24, 2012 (sol 137): this mosaic of images from Curiosity-Mastcam shows the rocks of Yellowknife Bay formation, that record superimposed ancient lake and stream deposits that offered past environmental conditions favorable for microbial life. Rocks here were exposed about 70 million years ago by removal of overlying layers due to erosion by the wind. Yellowknife Bay mudstone is thought to contain sediments transported by fluvial and deltaic processes from the crater rim area to the north.*

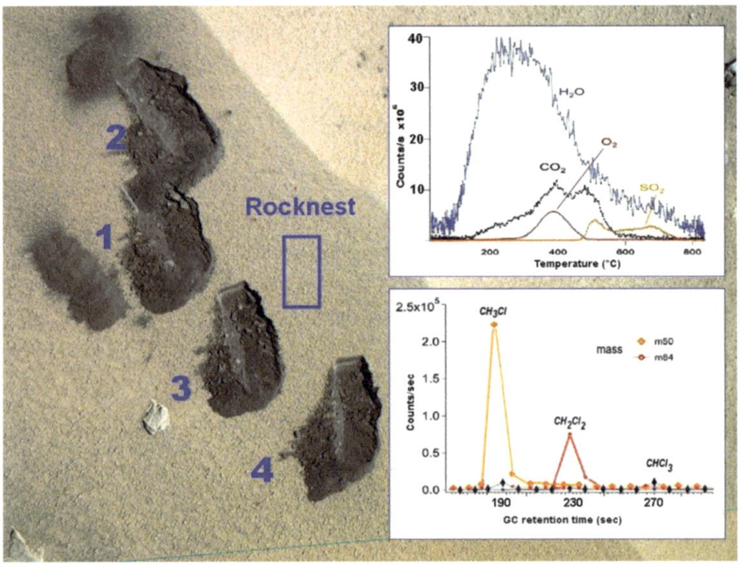


**Figure 31.**
*November 2012 (sol 93–117): results from analysis of Rocknest Aeolian deposit by SAM-Curiosity. On top the EGA evolved gases, on bottom some CGMS light chloro-derivatives.*

Cl, this emission is a suggestion of Ca $(ClO_4)_2$ (Calcium perchlorate), a salt that decomposes under heat just to this temperature.

Laboratory–based TGA (Thermal Gravimetrical Analysis, performed by the Author with a Perkin-Elmer TGA 7 instrument) on synthetic perchlorates shows clearly that the Calcium perchlorate starts to release molecular oxigen at 350°C, leaving a main residue of Calcium chloride ($CaCl_2$) (**Figure 32**). Therefore, after the discovery of perchlorate at high latitude by Phoenix, SAM demonstrated an occurrence of perchlorate also at equatorial latitude: so its occurrence also at mid-latitude (i.e. Viking landing sites) comes out strengthened.

Actually, between 200 and 500°C, the soil of Rocknest released water and two peaks of $CO_2$ (i.e. two releases at two different temperatures). The origin of this water and Carbon dioxide is doubtful. Being released at more than 200°C, the water cannot be free, but bound to soil minerals as water of crystallization. In addition, a lab simulation shows that the two peaks of $CO_2$ could arise from the thermal decomposition of Mg and Fe carbonate [54].

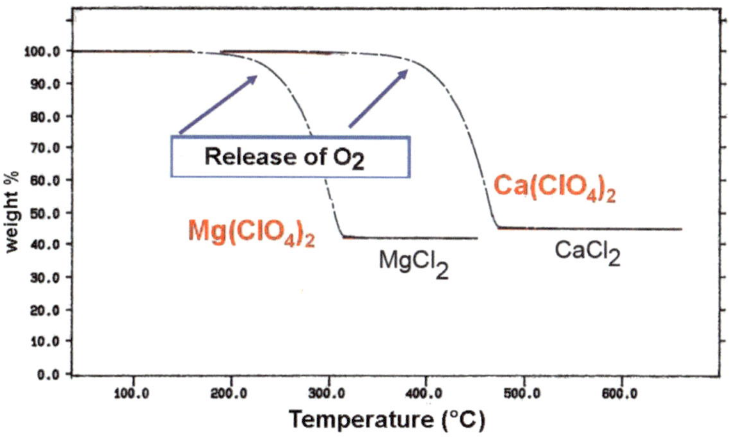


**Figure 32.**
*Thermal decomposition of perchlorates measured by the Author on a TGA (Thermal Gravimetric Analysis) instrument. Each perchlorate shows a specific temperature of decomposition with release of Oxygen.*

But alternative hypotheses could have been given. The water and carbon dioxide seen by SAM could be breakdown products of organic substances under the action of perchlorates. This claim results from another discovery of GCMS on board SAM: the detection of simple chlorinated molecules, such as $CH_3Cl$ and minor amount of $CH_2Cl_2$ and $CHCl_3$ [25]. At the end of February 2013 the SAM made a second series of analyses on a powdered sample of a sedimentary terrain named John Klein (**Figure 33**), located about 50 meters away from Rocknest, confirming results of first analysis, i.e. emission of $CO_2$ and $H_2O$, of $O_2$ over 250°C (probably generated by perchlorates dissociation), and release of $CH_3Cl$ + $CH_3Cl_2$ [55].

Therefore, the SAM and Viking GCMS results look strikingly similar, in the sense that a sufficient amount of perchlorates could mask occurrence of organics.

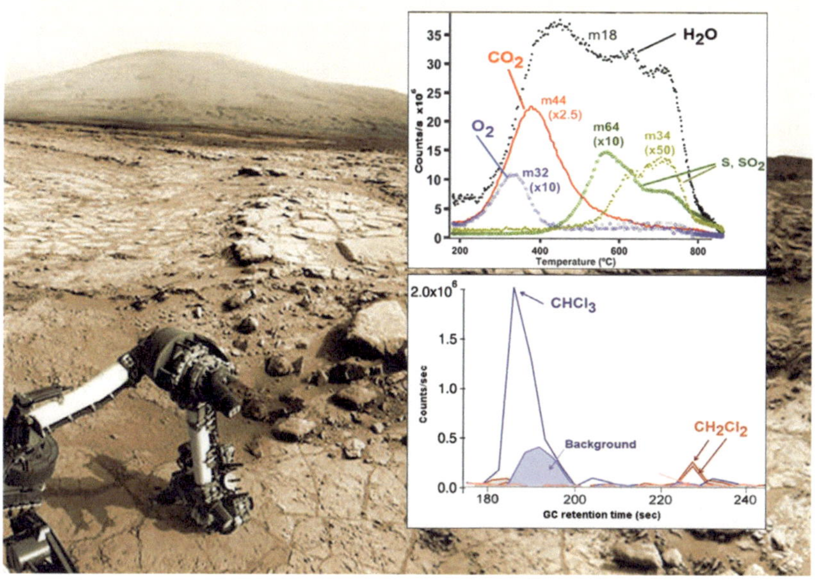


**Figure 33.**
*February 2013: results by SAM-Curiosity from a powdered material drilled into the John Klein sedimentary rock. On top the EGA evolved gases, on bottom some CGMS light chloro-derivatives.*

After many months of stop due to a serious pollution problem (see later), the SAM team started again its analytical work, on a soil sample of the site of Cumberland that was taken an year before, on May 2013 not far from John Klein. The SAM results were crucial [56]: aside from the usual light Chloro-derivatives, many chlorinated aromatics were detected [57] suggesting that they could be derived from organic molecules present in the mudstone (from bacteria or from a meteoric extract): between them also an abundant release (about 250 ppb) of Chloro-benzene was detected (**Figure 34**), so reaching for an other resemblance to the Viking results, in which Chloro-benzene (as mentioned before) was found after a recent accurate re-examination of the original GCMS data [29].

One of the most extraordinary SAM discovery was made at Pahrump Hills (at the base of~3.5-billion-year-old Murray mudstone), located at the lowermost portion of the Sharp Mons (Gale Crater central mound), about 6–7 km southwest of Yellowknife Bay. This ~3.5-billion-year-old Gale lake environment is expected to have been ideal settings for concentrating and preserving organic matter [58]. Two samples were drilled: Confidence Hills on sol 759 (24 Sep 2014), and Mojave on sol 882 (29 Jan 2015) (**Figure 35**). Confidence Hills soil was rich of hematite, Mojave soil was rich of jarosite, evidence of ancient passage of water. Because ultraviolet radiation and oxidizing compounds in the Martian soil would destroy any compounds exposed at the surface, Curiosity's scientists used a robotic drill to penetrate several centimetres into the mudstone.

To unlock organic molecules from the samples, the oven baked them to temperatures of between 600°C and 860°C and fed the resulting fumes to the Mass Spectrometer, which identified a welter of closely related organic signals reflecting dozens or hundreds of types of small carbon molecules, such as aromatic rings and short aliphatic chains [59]. Abundant sulfur-bearing carbon rings called thiophenes, were also detected and identified in the GC (**Figure 36**).

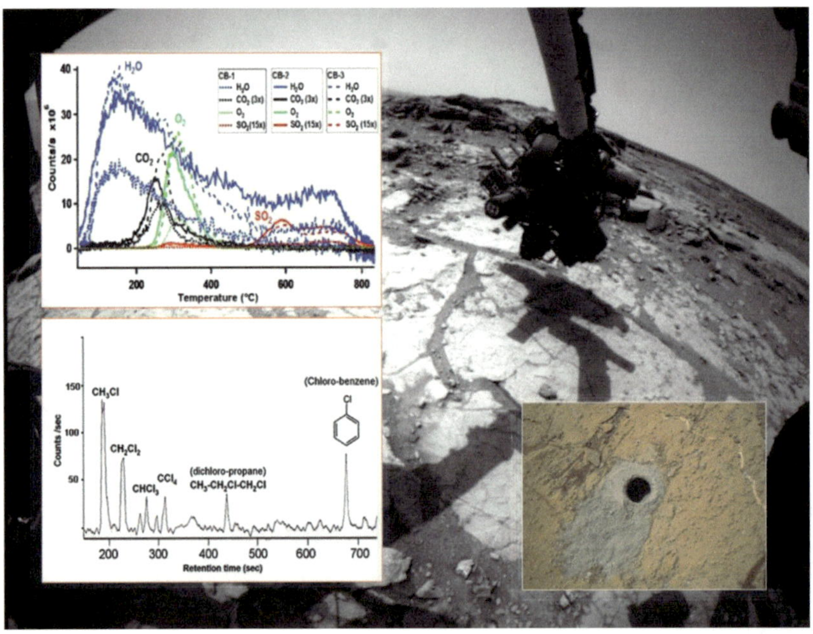


**Figure 34.**
*December 2014: results by SAM-Curiosity from a powdered material drilled into the Cumberland sedimentary rock. Very important (bottom), between the GCMS evolved chloro-derivatives, the presence of chloro-benzene, a byproduct certainly of Martian origin.*

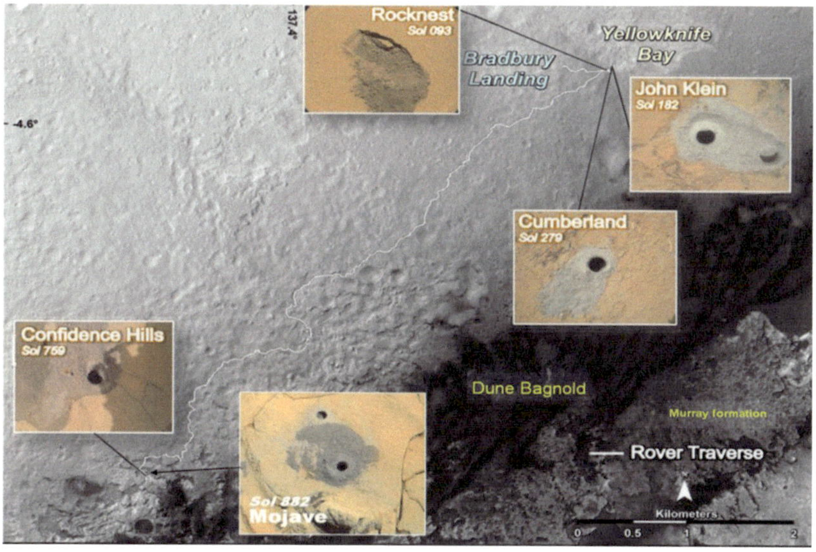


**Figure 35.**
*Confidence Hills and Mojave drill sites, at Pahrump Hills location, were SAM made the main discovery of possible keragenic material.*

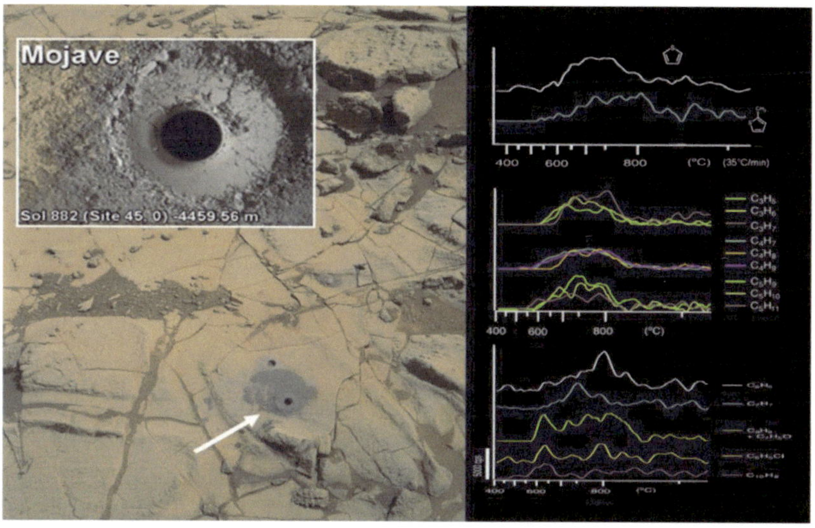


**Figure 36.**
*SAM-EGA evolved gases from Majave drilled material: sulfur-bearing carbon rings, short aliphatic chains and aromatic rings are typical decomposition products of keragenic material.*

The mass patterns looked like those generated on Earth by kerogen (aromatic rings, short aliphatic chains, sulphur containing molecules), a goopy high molecular material that is formed when geologic forces compress, during million of years, the ancient remains of algae and similar critters. Kerogen is sometimes found with sulfur, which helps preserve it across billions of years; the Curiosity scientists think the sulfur compounds in their samples also explain the longevity of the Mars compounds. At the moment, it is impossible to say whether ancient life explains the Martian organics. The signal, being found at the base of a lake 3,5 billion years old, when Mars environment was warm and wet, could be a potential catchment for the presence on Mars of archea bacteria in primordial epoch, possibly still present today where there are sources of liquid water over the surface (superficial melting of ices rich in salts) [60] or below the surface (sub-glacial lakes identified by radar tecniques) [61]. However, we must not forget that Carbon-rich meteorites and comets contain kerogenic like compounds, and constantly rain down on Mars ...

It's disappointing that we can't figure out where the carbon-rich large molecules came from. But digging a little deeper could find better-preserved molecules in Mars rocks, to determine whether these molecules came from space, from igneous rocks, from hydrothermal activity, or – the most exciting possibility – ancient Mars life. Europe's ExoMars rover, due for launch in 2022, will drill deeper than Curiosity, to soil depths better protected from radiation. But probably, detection of past life may ultimately take the precision analysis of labs on Earth, bringing samples back. Fortunately, NASA Perseverance rover, that was successful in landing inside the Jezero crater on February 18, 2021 (**Figures 37** and **38**), is set to collect some 30 rock cores for return to Earth in subsequent missions.

SAM is able to give a further usefull help to determine the nature and origin of the kerogenic materials discovered on Mars, by the so called 'wet chemistry' experiments.

In summary, if organic molecules cannot enter the GCMS because a low volatility or breaking dawn under heating, they can be "derivatized" before they're heated – meaning that they react with some chemicals in order to become more volatile – so that they can be analyzed at a lower temperature. This derivatization process uses special chemical reagents dissolved in suitable solvents, so this experiment is called "wet chemistry". As yet mentioned, SAM only has nine Inconel steel cups containing these derivatizing agents: 7 containing a derivatization-silanizing

**Figure 37.**
*The Jezero crater, an ancient lake where the rover Perseverance landed on February 18, 2021.*

**Figure 38.**
*February 21, 2021: Perseverance sees Jezero crater rim in 360° Mars panorama.*

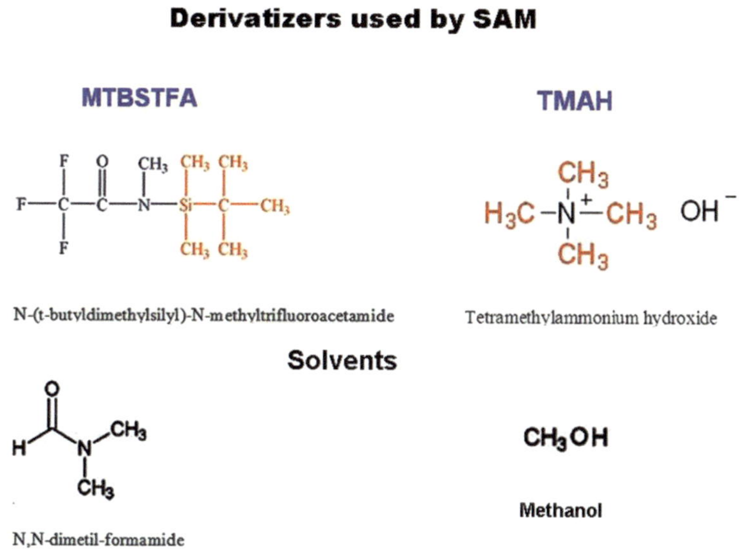


**Figure 39.**
*The two derivatizing reagents for 'wet chemistry' on board of SAM-Curiosity.*

compound named MTBSTFA (N-tert-butyldimethylsilyl- N-methyltrifluoroacetamide), 2 containing a thermochemolysis compound named TMAH (tetramethylammonium hydroxide) (**Figure 39**).

MTBSTFA is an organic compound containing Florine and Silicium, able to instantly replace active hydrogens on OH and $NH_2$ (carboxylic acid, amine, amino-acid) with a N-tert-butyldimethylsilyl group (**Figure 40**): this non-polar moiety increases the volatility of the original compound by removing its polar nature, resulting in a much lower temperature needed for a GCMS analysis [62]. Due to the limited number of cups for 'wet chemistry', these kinds of experiments were obviously saved for only the most interesting rock samples. But an incredible accident caused the first wet chemistry trial to be postponed for six years. During the

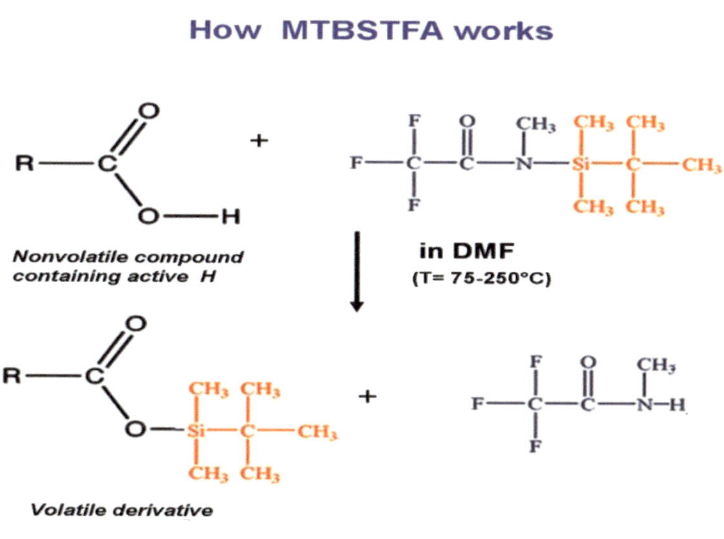


**Figure 40.**
*The MTBSTFA chemical mechanism of derivatization, called silanization.*

examination of the results the SAM obtained on Rocknest and on John Klein, the SAM team discovered that a vial of MTBSTFA was broken, so polluting all the analytical system.

The problem was that MTBSTFA, being itself an organic compound, reacts under heat with perchlorates, giving the same kind of light chloro-derivatives ($CH_3Cl$ and $CH_2Cl_2$) found by the SAM on the Martian samples! From here a dreadful doubt that the origin of the 'positive' results obtained so far by the SAM could be 'terrestrial' and not Martian [25]. More than a year was needed to clean the system, during which a sample from Cumberland remained stored inside SAM, waiting the right moment to be analyzed. The sample remained 1280 sols (!) in contact with MTBSTFA vapors, a situation that also provided an opportunity [63]: baking the sample up to ~900°C to verify if some reaction between MTBSTFA and Martian soil had happened. This so called 'opportunistic derivatization' was a success, because the GC–MS detected interesting compound such as Chlorobenzene, Thiophene, light Chloro-derivatives and many other unknown compounds (**Figure 41**).

Lab tests demonstrated that Chloro-benzene, an organic compound containing 6 Carbon atoms, could not be formed from the heating of MTBSTFA in presence of perchlorates [64] but only when various types of Martian organic materials are pyrolyzed (i.e. heated at high temperature) in presence of Chlorine source.

The first 'complete' wet chemistry experiment was made on December 19, 2017. The target was a Ogunquit Beach (OB) sand sample from the Bagnold dune field, chosen being easy to manipulate after months of trouble due recurrent problems with the drill feed mechanism (**Figure 42**).

About 45 mg of the Ogunquit Beach sand were added to one of the MTBSTFA cups and the mixture was heated up to 900°C. Reactions clearly occurred and produced derivatized compounds. GCMS results showed the detection of derivatized benzoic acid as well as excess, unreacted MTBSTFA. However, no amino acids or fatty acids were detected [65].

During the following months engineers found a way to fix the drill problem. So the next step was to perform wet chemistry experiments on drilled clay deeper samples, as these phyllosilicate-rich minerals are known to preserve organic matter exceptionally well. A unusual Mn- and P-rich clay-bearing unit named Glen Torridon (GT) (**Figure 43**) was discovered in the foothills of Mount Sharp by CRISM spectrometer aboard the Mars Reconnaissance Orbiter (strong absorptions at 2.24 and 2.29 μm. EGA analyses in many locations inside Glen Torridon showed emission of free and strongly linked water and, surprisingly, absence of emission of

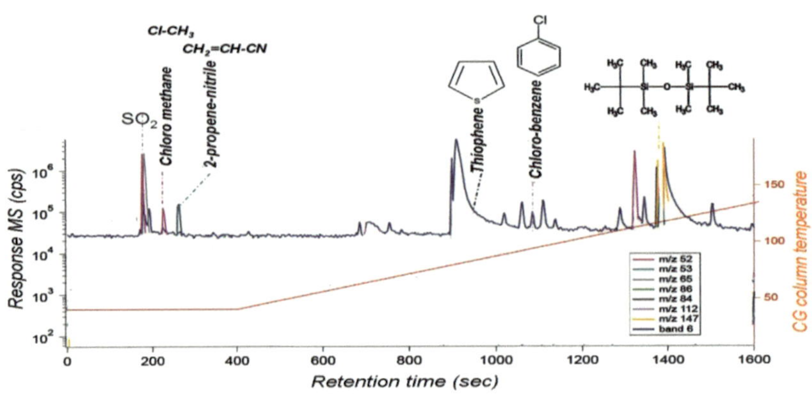


**Figure 41.**
*The GCMS result of the so called 'opportunistic derivatization', performed on Cumberland material, that stayed in touch for months with MTBSTFA vapors accidentally leaked from a broken cup.*

**Figure 42.**
*Ogunquit Beach where SAM performed the first complete derivatization with MTBSTFA.*

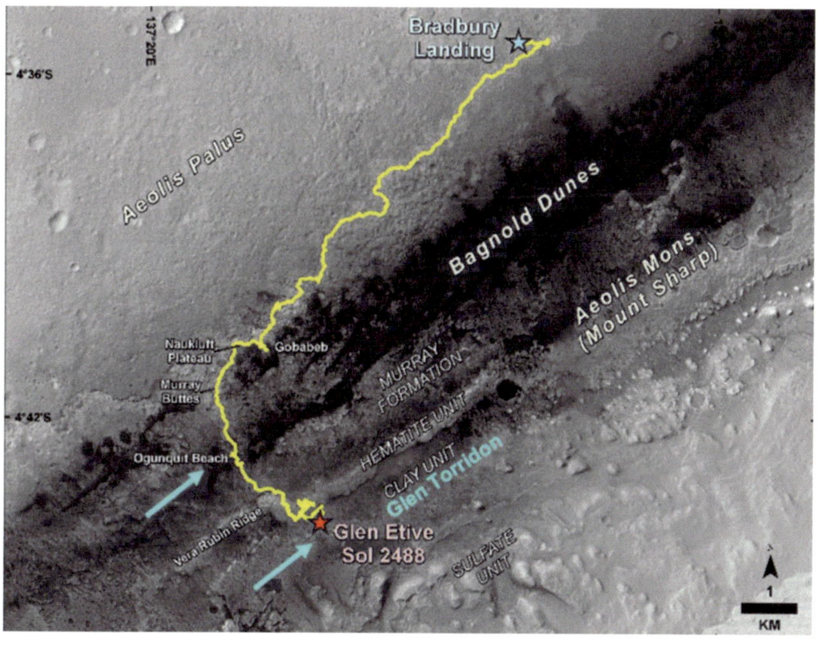


**Figure 43.**
*The clay-bearing unit named Glen Torridon where SAM performed also the first derivatization with TMAH (called thermochemolysis).*

$O_2$ from perchlorates [66]: a promising situation in order to search for organic compounds. SAM activities in Glen Torridon included an EGA/GCMS analysis, a new MTBSTFA derivatization experiment, followed by the first TMAH experiment. On Sept. 24, 2019 (sol 2536) the rover placed in the SAM the powderized drilled sample from GT- Glen Etive 2 site (**Figure 44**).

EGA/GCMS detected an abundance of S-bearing organic compounds, including aliphatic and aromatic compounds: dimethylsulfide, thiophene, and likely ethanethiol and dithiapentane. EGA also indicated results within the medium to high molecular weight ranges of masses, suggesting the presence of a complex mixture of compounds.

**Figure 44.**
*The Glen-Etive-2 site (Glen Torridon clay unit) where SAM-EGA discovered high molecular weight carbon molecules, possibly resulting from the breakdown of keragenic material.*

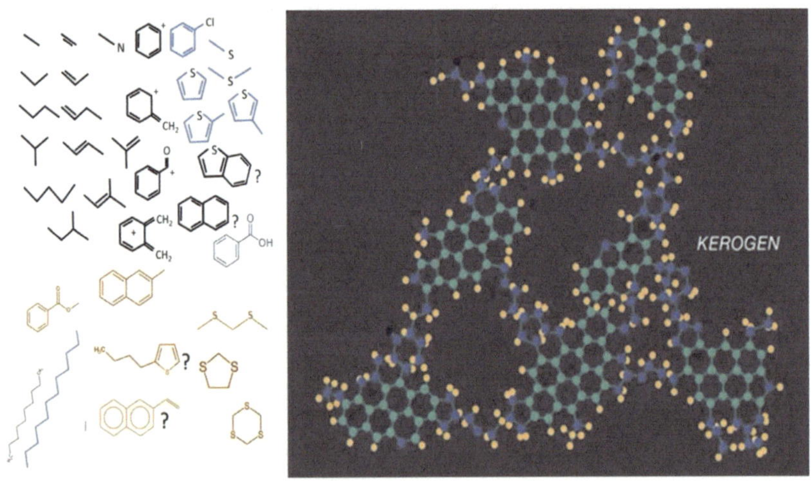


**Figure 45.**
*SAM EGA + CGMS have revealed a range of organic fragments detected above 500°C, indicating the presence of recalcitrant organic matter (e.g. macromolecles). Compounds in blue were only detected via GCMS. Compound in yellow are from Glen Torridon.*

The diversity of aromatics seems consistent with recalcitrant organic materials such as kerogenic-type macromolecules (remembering what was found at Cumberland and Mojeve five years before) (**Figure 45**).

MTBSTFA experiment showed the highest abundance of sulfur-bearing organics ever measured by the SAM instrument and a wide range of aromatic organic molecules including methylated polycyclic aromatic hydrocarbons (methyl-naphthalene), a potential methylated ester carboxylic acid (benzoic acid) and Benzothiophene, all detected for the first time on Mars. However, no amino acids or fatty acids have been identified [67].

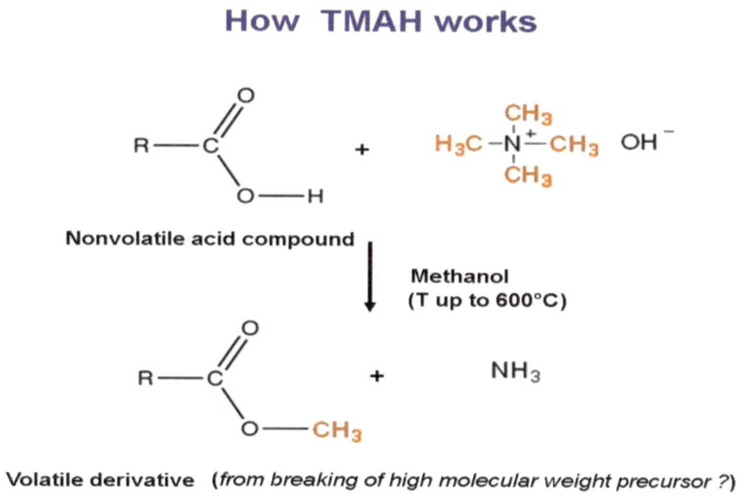


**Figure 46.**
*The TMAH chemical mechanism of derivatization, called thermochemolysis.*

The possible discovery of high molecular molecules, was the long awaited reason for the first in situ TMAH wet chemistry experiment (**Figure 46**), the so called thermochemolysis, performed by hearing a sample of Martian soil in contact with one of the two cups onboard of SAM, containing tetramethylammonium hydroxide (TMAH), dissolved in Methanol [68].

This strongly alkaline reagent causes hydrolysis and methylation of -OH, -O-, -NH, and -SH groups bonds and, upon heating, thermal bond breakage (of big molecules) also enuses. Volatile products of thermochemolysis were directly analyzed by the mass spectrometer (EGA) or trapped and analyzed with gas chromatography mass spectrometry (EGA-GCMS). This amazing experiment was successfully executed in September 2020 at the Mary Anning (MA) drill site (**Figure 47**) in the Glen Torridon region [69, 70].

'Bands' of masses grouped together and having mass-to-charge (m/z) 190 to 485, represent high molecular weight molecules detected by the SAM-MS. These

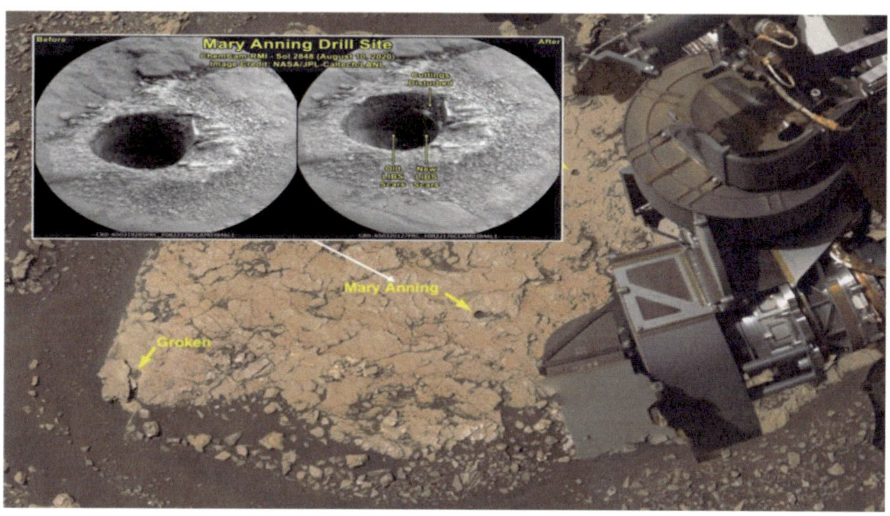


**Figure 47.**
*The Mary Anning drill site (Glen Torridon clay unit) where the first TMAH derivatization was perfomed on September 2020.*

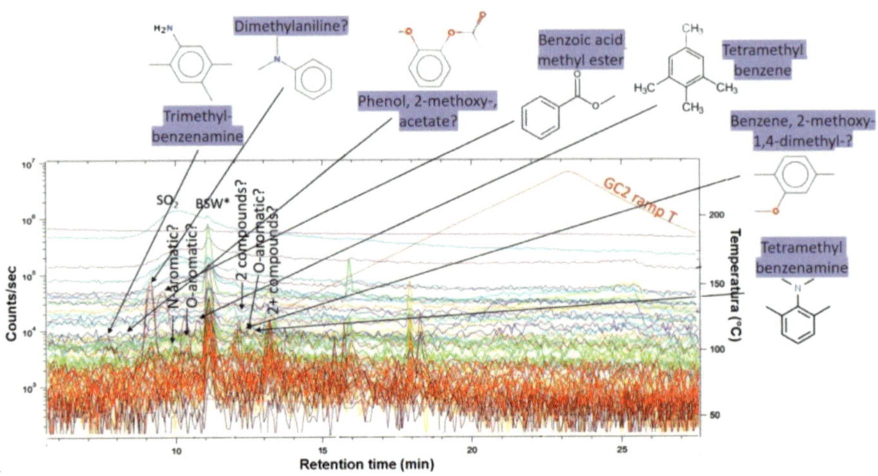


**Figure 48.**
*The preliminary astonishing results of the TMAH thermochemolysis on the Martian soil drilled at Mary Anning. Separation was performed using a gas-chromatographic column (GC2) able to separated molecules with more than 15 Carbon atom. The red continuous line shows the ramp of temperature (adapted from A. Williams, fall 2020 AGU meeting).*

data may indicate that large, complex molecules were present. A variety of methylated, oxygen-, sulfur-, or nitrogen-bearing aromatic organics were detected in GCMS and/or EGA data. The presence of methylated single and double ring aromatics (included benzene, toluene, trimethyl- and tetramethyl-benzene, naphthalene, and methylnaphthalene) suggests that these organics might derive from a macromolecular source that was cleaved and methylated by TMAH thermochemolysis (**Figure 48**).

Examples of the organics detected in GCMS only include pentamethyl-benzene, benzoic acid methyl ester, dimethyl-, trimethyl-, and tetrame-thyl-benzenamine, dihydronaphthalene, 2-butyl-thio-phene, and benzothiophene. Pentamethylbenzene may be part of a multi-methylated benzene suite. The benzoic acid methyl ester reflects the reaction of TMAH methylating benzoic acid of indeterminate source. The multi-methylated benzenamine suite is also of indeterminate source. A non biological source of these organics on the surface of Mars could be the impact of meteoritic material. Several similar organics were indeed identified applying the TMAH thermochemolysis benchtop experiment to the Murchison meteorite, including toluene, trimethylbenzene, methylnaphthalene, 2-butyl-thiophene, and benzothiophene. Amines and amides are not prevalent in pyrolyzed Murchison material and benzenamines are also not generated during TMAH thermochemolysis of Murchison, so the origin of these amines remains at least problematic.

## 5. Conclusions

The discovery of organic molecules on Mars is a necessary but not sufficient condition for the demonstration of some present or past form of martian bacterial life. Comets and the carbonaceous chondrites are indeed copious sources of organic material. But the positive results of the 'famous' Viking LR experiment are intriguing, even because it is important taking in account that Viking LR was performed at low temperature, a condition where possible perchlorates do not show any oxidant effects.

## Author details

Cesare Guaita
GAT/Milano Planetarium, Tradate, Italy

*Address all correspondence to: c.guaita@libero.it

## References

[1] Jones BW. Mars before the space age. International Journal of Astrobiology. 2008;7 (2):143-155

[2] *Sheehan W. The Planet Mars: A History of Observation and Discovery. The University of Arizona Press; 1996*

[3] Schiaparelli GV. The planet Mars. A&A. 1894;13:635-723

[4] Herr C, Pimentel C. Infrared Absorptions near Three Microns Recorded over the Polar Cap of Mars. Science. 1969;166:496-499

[5] Soffen GA. The Viking Project. JGR. 1977;82:3959-3970

[6] Biemann K et al. The Search of Organic Substances and Inorganic Volatiles Compounds in the Surface of Mars. JGR. 1977;82:4641-4668

[7] Klein H. et al. The Search for extant life on Mars. In: Mars. Univ. Arizona Press;1992.p.1221-33

[8] Sharp ZD et al. Chlorine isotope homogeneity of the mantle, crust and carbonaceous chondrites. Nature.2007; 446:1062-1065

[9] Bains W. Martian Methyl Chloride. A lesson in incertainty. 2013; arXiv: 1304.4429

[10] Steele A et al. A Reduced Organic Carbon Component in Martian Basalts. Science. 2012;337: 221-215

[11] *Thomas-Keprta KL et al.* Origins of magnetite nanocrystals in Martian meteorite ALH84001. Geochimica et Cosmochimica Acta. *2009;*73: 6631-6677

[12] *McKay D et al.* Search for Past Life on Mars: Possible Relic Biogenic Activity in Martian Meteorite AL84001. Science. *1996;273:924–929*

[13] Uwins JR et al. Novel nano-organisms from Australian sandstones. American Mineralogist. 1998;83: 1541-1550

[14] Oyama V, Bendahi B. The Viking Gas Exchange Experiment Results From Cryse and Utopia Surface Samples. JGR. 1977;82:4669-4676

[15] Horowitz NH et al. Viking at Mars: The Carbon Assimilation Experiment. JGR. 1977;82:4659-4667

[16] Levin G, Straat P. Recent Results From the Viking Labeled Release Experiment on Mars. JGR. 1977;82: 4663-4667

[17] Miller JD et al. Periodic analysis of the Viking lander Labeled Release experiment, In: Proceedings SPIE 4495, Instruments, Methods, and Missions for Astrobiology IV; 5 February 2002; p.96-107

[18] Bianciardi G et al. Complexity Analysis of the Viking Labeled Release Experiment. IJAAS. 2012;13:14-26

[19] Levin G, Straat P. Viking Labeled Release Experiment:Interim Results. Science. 1976;194:1322-1328

[20] Klein H. The Viking Biological Investigation: General Aspect. JGR. 1977; 82: 4677-4680

[21] Yen AS et al. Evidence of the great reactivity of the Martian soil is due to superoxide ions. Science. 2000; 289: 1909-1912

[22] Ponnamperuma C. et al. Possible surface reaction on Mars: implication for Viking biology results. Science. 1977;197: 455-457

[23] Cotton FA et al. Basic Inorganic Chemistry. Wiley;1995. pp 435-443

[24] Hecht M et al. Detection of Perchlorate and the Soluble Chemistry of Martian Soil at the Phoenix Lander Site. Science. 2009;325:64-67

[25] Glavin DP et al. Evidence for perchlorates and the origin of chlorinated hydrocarbons detected by SAM at the Rocknest aeolian deposit in Gale Crate. JGR Planets. 2013;118:1-19

[26] Kounaves S P et al. Evidence of Martian perchlorate, chlorate, and nitrate in Mars meteorite EETA79001: Implications for oxidants and organics. Icarus. 2014;229:206–213

[27] Navarro-Gonzales R et al. Reanalysis of the Viking results suggests perchlorate and organics at mid-latitudes on Mars. JGR. 2010;115 (E12): 1-11

[28] Steininger H et al. Influence of magnesium perchlorate on the pyrolysis of organic compounds in Martian soil analogs. Planetary and Space Science. 2012;71:9-17

[29] Guzmann M et al. Identification of Chlorobenzene in the Viking Gas Chromatograph-Mass Spectrometer Data Sets. JGR Planets. 2018;123: 1674-1683

[30] Clark BC et al. Chemical composition of Martian fines. JGR. 1982; 87(B12):10059-10067

[31] Biemann K, Bada JL. Comment on "Reanalisys of the Viking results suggests perchlorate and organics at midlatitude on Mars". JGR. 2011;116: E12001-E12005

[32] Foley CN et al. Final chemical results from the Mars Pathfinder alpha proton X-ray spectrometer. JGR. 2003; 108(E12):8096-8115

[33] Gellert R et al. The Alpha-Particle-X-ray-Spectrometer (APXS) for the Mars Science Laboratory (MSL) Rover Mission. LPSC. 2009;40: 2364-2365

[34] Rieder R et al. Chemistry of rocks and soils at Meridiani Planum from the Alpha Particle X-ray Spectrometer. Science. 2004;306:1746-1749

[35] Keller JM et al. Equatorial and midlatitude distribution of chlorine measured by Mars Odyssey GRS. JGR. 2006;111:1-18

[36] Catling, D C et al. Atmospheric origins of perchlorate on Mars and in the Atacama, JGR. 2010;115(E00E11):1-15

[37] Carrier B L, Kounaves SP. The origins of perchlorate in the Martian soil. Geophys. Res. Lett. 2015;42: 3739-3745

[38] Wilson E et al. Perchlorate formation on Mars through surface radiolysis-initiated atmospheric chemistry: A potential mechanism. JGR Planets. 2016;121:1472-1487

[39] Schuttlefield JD et al. Photooxidation of Chloride by Oxide Minerals: Implications for Perchlorate on Mars. J. Am. Chem. Soc. 2011;133: 17521-17523

[40] Marvin GG, Woolaver LB. Thermal Decomposition of Perchlorates. Ind. Eng. Chem. Anal. *1945;*17 (8):474-476

[41] Baucon A et al. Ichnofossils, Cracks or Crystals? A Test for Biogenicity of Stick-Like Structures from Vera Rubin Ridge, Mars. Geosciences. 2020;10:39-57

[42] Rizzo V, Cantarano N. Structural parallels between terrestrial microbialites and Martian sediments: are all cases of 'Pareidolia'?. International Journal of Astrobiology. 2017;16 (4):297-316

[43] Mahaffy PR et al. The Sample Analysis at mars Investigation and

Instrument Suite. Space Sci. Rev. 2012; 170:401-478

[44] Tarsitano CG, Webster C.R. Multilaser Herriott cell for planetary tunable laser spectrometers. Appl. Opt. 2007;46(28):6923-6935

[45] Mumma MJ et al. Strong Release of Methane on Mars Northern Summer 2003. Science. 2009;323: 1041-1045

[46] Formisano V et al. Detection of Methane in the Atmosphere of Mars. Science. 2004;306:1758-1761

[47] Webster C et al. Background levels of methane in Mars atmosphere show strong seasonal variations. Science. 2018;360:1093-1096

[48] Gillen E. Statistical analysis of Curiosity data shows no evidence for a strong seasonal. Icarus, 2020;326:113407

[49] Garcia JP et al. Comparing MSL Curiosity Rover TLS-SAM Methane Measurements With Mars Regional Atmospheric Modeling System Atmospheric Transport Experiments. JGR Planets. 2019;124:2141-2167

[50] *Giurann M et al. Independent confirmation of a methane spike on Mars and a source region east of Gale Crater.* Nature Geoscience. *2019;12 (5):326-332*

[51] Korablev O et al. No detection of methane on Mars from early ExoMars Trace Gas Orbiter observations. Nature. 2019;568:517-520

[52] Gellert R. et al. Chemistry of Rocks and Soils in Gusev Crater from the Alpha particle X-ray Spectrometer. Science. 2004;305:829-832

[53] Blake DF et al. Curiosity at Gale Crater, Mars: Characterization and Analysis of the Rocknest. Science (special issue). 2013;341:1239505/ 1-1239505/7

[54] Leshin LA et al. Volatile, Isotope, and Organic Analysis of Martian Fines with the Mars Curiosity Rover. Science (special issue). 2013;341:1238937/ 1-1238937/9

[55] Ming DW et al. Volatile and Organic Compositions of sedimentary Rocks in Yelloknife Bay, Gale Crater, Mars. Science (Special issue). 2014;343: 1245267/1-1245267/9

[56] Freissinet C et al. Organic molecules in the Sheephed Mudstone, Gale crater, Mars. JGR Planets. 2015;120:495-514

[57] Szopa C et al. First Detections of Dichlorobenzene Isomers and Trichloromethylpropane from Organic Matter Indigenous to Mars Mudstone in Gale Crater, Mars. Astrobiology. 2020; 20:292 -305

[58] Summons RE et al. Preservation of Martian Organic and Environmental Records: Final Report of the Mars Biosignature Working Group. Astrobiology. 2011;11:157-181

[59] Eigenbrode J et al. Organic matter preserved in 3-billion-year-old mudstones at Gale crater, Mars. Science. 2018;360:1096-1101

[60] Ojia L et al. Spectral evidence for hydrated salts in recurring slope lineae on Mars. Nature Geoscience. 2015;8: 829-832

[61] Lauro SE et al. Multiple subglacial water bodies below the south pole of Mars unveiled by new MARSIS data. Nature Astronomy. 2021;5:63-70

[62] Kataoka H. Derivatization reactions for the determination of amines by gas chromatography and their applications in environmental analysis. J. of Cromarography. 1996;733:19-34

[63] Freissinet C et al. Opportunistic derivatization to investigate organics in

a martian mudstone. LPSC. 2017;58: 2886-2887

[64] Miller KE et al. Potential precursor compounds for chlorohydrocarbons detected in Gale Crater, Mars, by the SAM instrument suite on the Curiosity Rover. JGR Planets. 2016;121:296-306

[65] Millan M et al. Optimization of the sample analysis at Mars wet chemistry experiment for the detection of organics in Glen Torridon. LPSC. 2020;51: 1897-1898

[66] Sutter B et al.Geological processes along the Glen Torridon. LPSC. 2020;51: poster

[67] Millan M. et al. Organic molecules revealed in Glen Torridon by SAM instrument. LPSC. 2021;52: 2039-2040

[68] Williams JA et al, Recovery of Fatty Acids from Mineralogic Mars Analogs by TMAH Thermochemolysis for the Sample Analysis at Mars Wet Chemistry Experiment., Astrobiology. 2019;19 (4): 522-546

[69] Williams JA et al. Organic matter hetereogeneity in Mary Anning/Groken drill site, Gale crater, Mars. LPSC. 2021; 52:1638

[70] Williams AJ et al. Organic molecules detected with the first TMAH wet chemistry experiment, Gale crater, Mars. LPSC. 2021;52:1763

Chapter 7

# The Challenge of Controlling a Small Mars Plane

*Seiki Chiba and Mikio Waki*


## Abstract

Dielectric elastomers (DEs) are lightweight and high-power, making them ideal for power control in a planetary exploration spacecraft. In this chapter, we will discuss the control of an exploration airplane exploring the surface of Mars using DEs. This airplane requires lightweight and powerful actuators to fly in the rare Martian atmosphere. DEs are a possible candidate for use as actuator controlling the airplane since they have high power, and high efficiency. A structural model of a wing having a control surface, a DE, and a linkage was built and a wind tunnel test of a control surface actuation using a DE actuator was carried out.


**Keywords:** mars airplane, control surface, dielectric elastomer, artificial muscle, actuator, high efficiency, high power, power control of planetary spacecraft, SWCNT

## 1. Introduction

Dielectric elastomer actuators (DEAs) are currently used in a variety of applications such as robots and medical devices. Since a DEA is very light and capable of high output, it is expected to be able to control the output of the planetary exploration ship and the solar panels loaded on exploration ships.

We are developing a DEA related to the output and control of Mars probes that observe the surface of Mars. Mars Airplane is a new Mars observation platform that enables a wide range of observations from low altitudes. Since 2010, the Japanese Mars spacecraft working group has been working on the conceptual design of Mars airplanes and various basic researches [1–5]. Mars exploration is performed with a weight of 6 kg, a wing width of 2.4 m, and a maximum cruising speed of 70 m/s. We are developing such a machine, as shown in **Figure 1** [6].

We would like to take this airplane down from the Mars exploration spacecraft with a parachute, disconnect the parachute at a high altitude (30 km) on Mars, control its flight remotely, and then gradually lower the altitude to fly a distance of about 300 km. Therefore, it is necessary to avoid high mountains on flights after the middle stage.

This Martian plane can obtain more detailed data than satellites and can observe a wider range than rovers. Also, one of the unique features of the Mars plane is to observe the formations of the canyons. Satellites cannot see the formation from the sky, and rovers cannot approach them.

Flight exploration by Martian planes has some difficulties, as detailed below:

These planes are not mass-produced and are very expensive due to their special payload and avionics for academic research. In addition to this, Mars planes must be

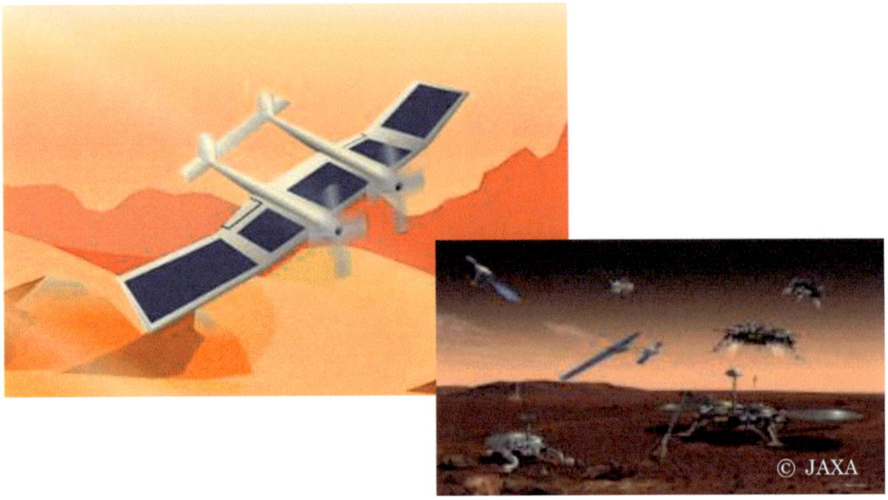


**Figure 1.**
*Image of the Mars exploration airplane.*

lightweight to fly in the thin atmosphere of Mars. It also has to be fairly lightweight to carry this spacecraft to Mars as well as to even be transported to Mars itself.

DEAs have the potential to be used as actuators for control surfaces (i.e. ailerons, rudder, and elevator) and as a propeller for the Mars airplane, since it is light and has high output and high efficiency. Another advantage of the DEA is that it is linearly driven, making it less susceptible to dust. This research investigated the feasibility of the DEA for the application of control surfaces (i.e., ailerons, rudder, and elevators) on the Mars airplane. A structural model of a wing having the control surface, the DEA, and a linkage was built, and a wind tunnel test of a control surface actuation using a DE actuator was carried out to investigate the feasibility of the DE actuators for the Mars airplane.

The results obtained in this study will be useful not only for the development of Mars exploration airplanes, but also for the structure and aerodynamic design of lightweight airplanes where large aerodynamic deformation is expected. The study also provides valuable examples of some of the expensive custom-made airplanes for academic research, airplanes that are not capable of many flight tests, and airplane development processes for which conducting flight tests are difficult.

## 2. Background of DE

To date, various types of soft actuators have been studied, and many functions desirable for different devices have been studied [7–31]. An especially attractive soft actuator is the dielectric elastomer (DE). DEs began to be studied in 1991 by R. Pelrine, S. Chiba et al. [16].

The basic element of a DE is a very simple structure comprised of a thin elastomer sandwiched by stretchable and flexible electrodes (see **Figure 2**) [24]. When a voltage difference is applied between the electrodes, they are attracted to each other by Coulomb forces leading to a thickness-wise contraction and plane-wise expansion of the elastomer. The typical thickness of the elastomers is about 500 microns to 1 mm. The electrode uses carbon black, CNT, or nano-sized metal. At the material level, the DE actuator has a fast speed of response (over 100,000 Hz), with a high strain rate (up to 680%), as shown in **Figure 3**, a high pressure, and a

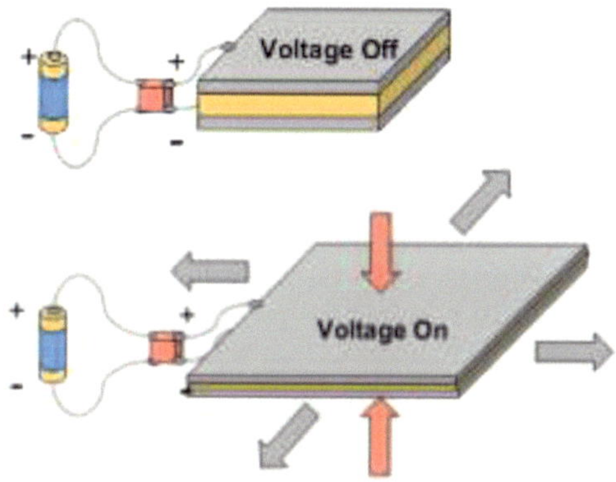


**Figure 2.**
*Principle of operation of DEs.*

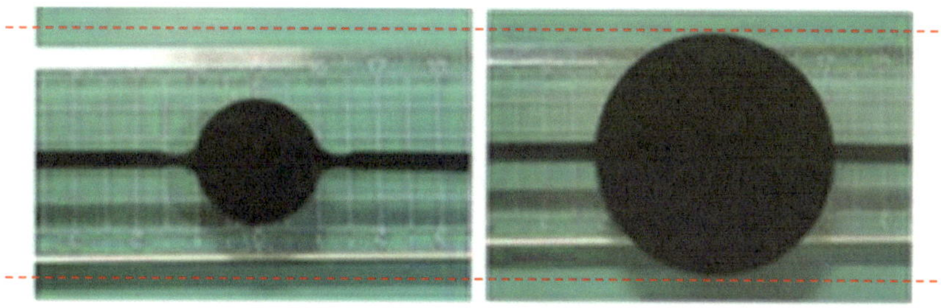

**Figure 3.**
*Expanding Circular Actuator up to 680.*

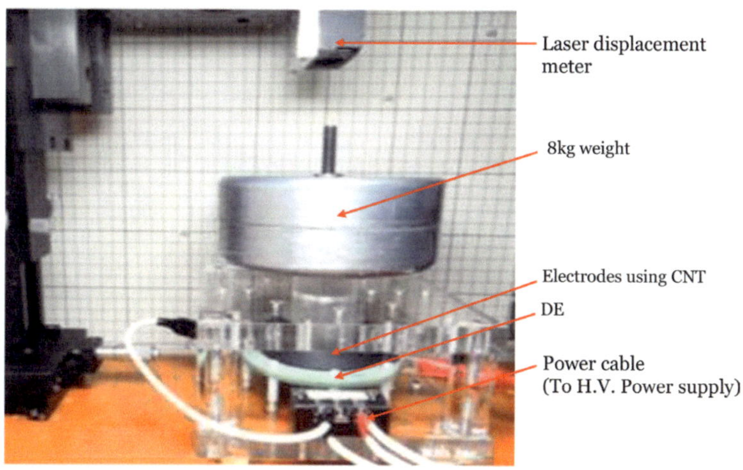


**Figure 4.**
*A DE lifted a 8.0 kgf weight using SWCNTs (ZEONANO®-SG101) (ZEONANO®-SG101 is a single-wall carbon nanotube synthesized by the Super-Growth method. (Diaphragm type DE actuator having a diameter of 8 cm)).*

power density of 1 W/g [32–34]. DEs can also be used for pressure-sensors and 3D position-sensors.

As shown in **Figure 4**, recently, DE actuators having only 0.15 g of DE material have been able to lift a weight of 8 kgf easily using the single wall carbon nano tube (SWCNT) electrodes (ZEONANO®-SG101) [33]. With 0.15 g of DEs, it is possible to lift an 8 kg weight by 1 mm or more. Its operating speed is 88 ms.

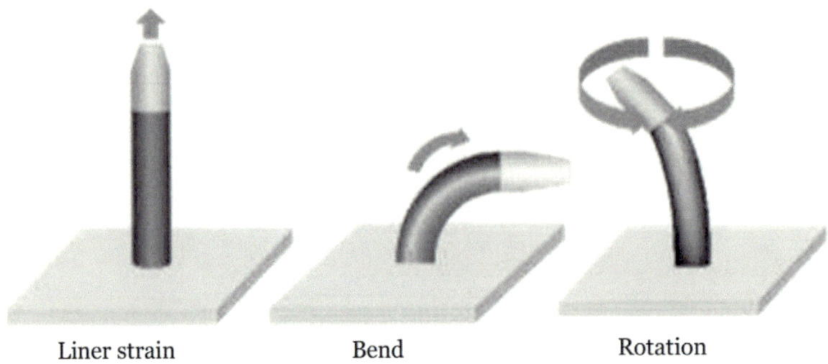


**Figure 5.**
*Roll actuator having 3-DOF.*

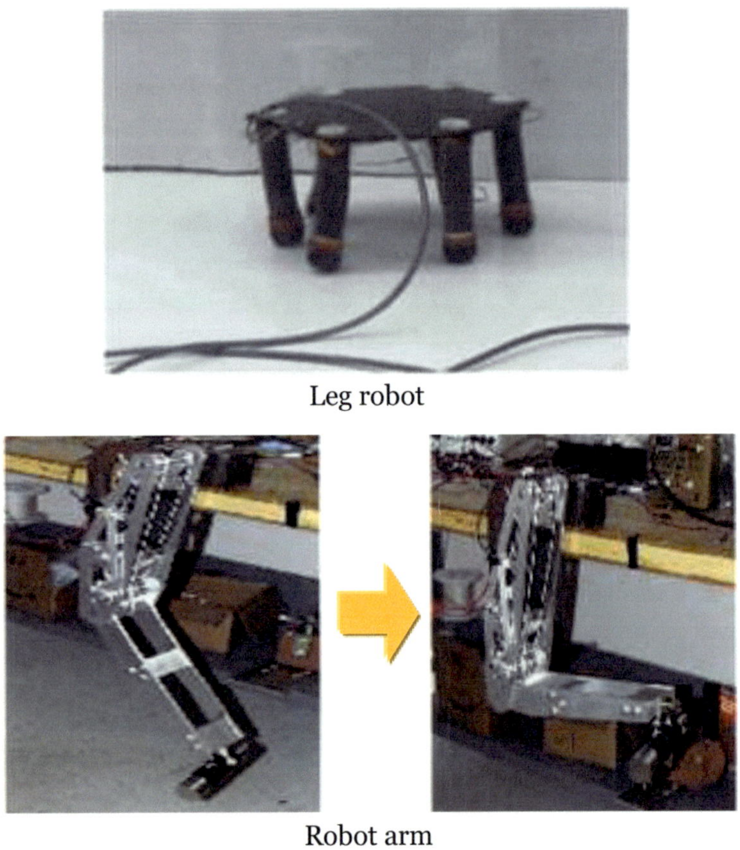


**Figure 6.**
*Biologically inspired robots powered by DE rolls.*

When the DE sheet is rolled, it becomes an actuator that looks like a human muscle. **Figure 5** shows the roll actuators with 3-DOF [34]. As shown in **Figure 6**, a DE can be the arm or leg of a robot. Using five of them, we created a robot that can moves around the surface of Mars [34], so it enables sideways stepping like a crab without turning around, when it collides with wall.

These roll-type actuators seem to be ideal for moving the antenna or solar panel of a space craft to the correct position and as an actuator for a working robot arm on it. For example, a robot arm is attached to the Japanese experimental module

**Figure 7.**
*International Space Station: Kibo (Japanese experiment module) has a robot arm.*

Kibo of the International Space Station (see **Figure 7**). It is possible to use DEs as the drive source.

## 3. Experimental procedure

In order to verify the possibility of using a DEA as a surface control actuator for the Mars exploration airplane, a wind tunnel was used to operate with the DEA while receiving wind.

### 3.1 Experimental setup

Using a continuously circulating low-speed wind tunnel (see **Figure 8**) owned by JAXA (Japan Aerospace Exploration Agency), we conducted a verification

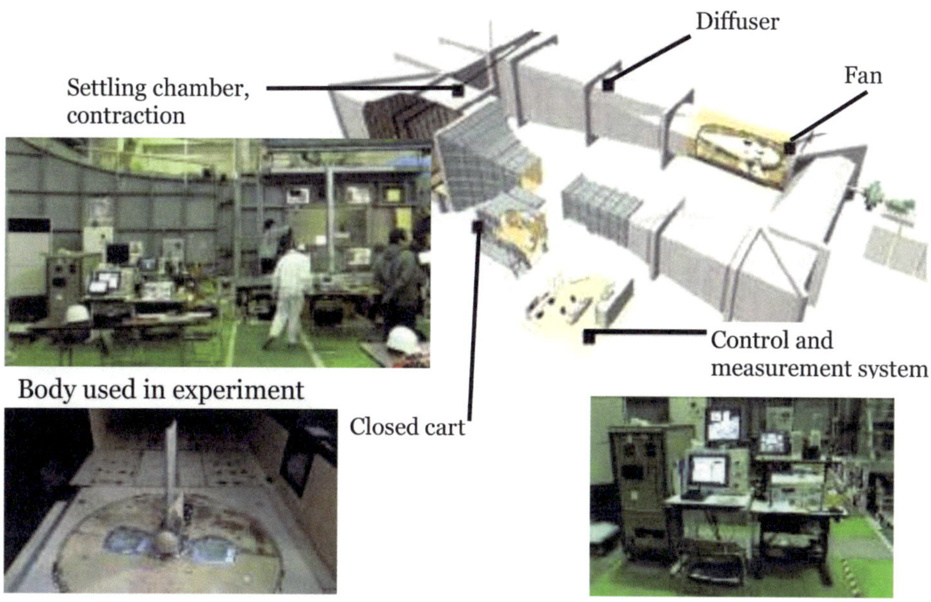


**Figure 8.**
*2 m x 2 m Low-speed wind tunnel.*

experiment to verify whether the DEA actuator could be used as a steering actuator under the wind received during flight. The wind tunnel used in this experiment has a measurement section of 2 m x 2 m and a maximum wind speed of 67 m / s (up to 60 m/s during continuous operation using the model).

In the preliminary study, the wind speed was set from 0 m/s to 40 m/s in order to match the actual driving conditions as much as possible. In addition, it was decided to observe the aileron driving state by changing the angle of attack from 0° to ±10° at 5° intervals at each wind speed (**Figure 9**).

The structural model of the wing used in the experiment was shaped vertically so that the surface control actuator (in this case, an aileron-like structure) can be driven by the DEA in a limited space in the wind tunnel. The body is shown in **Figure 10** [6].

The dimensions of the wings used in this experiment were 168.5 mm wide, 633 mm high and 25 mm thick, and the dimensions of the ailerons were 78.5 mm wide and 633 mm high. The wings and ailerons were made using polycarbonate resin for the frame and ABS resin for the exterior. The bottom of the model was 125 mm in diameter and 757.5 mm in length. The tip and tail are made of ABS resin,

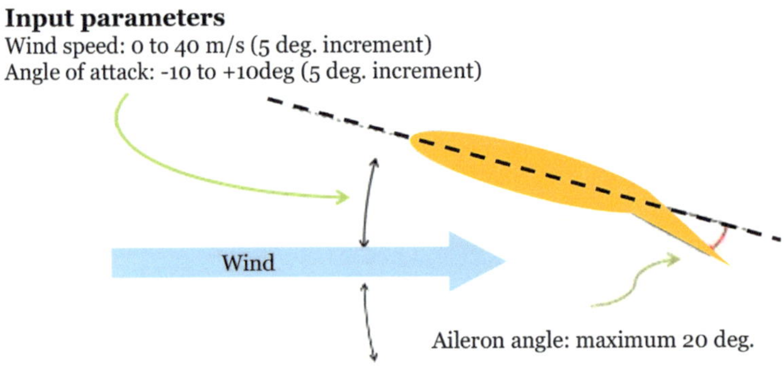


**Figure 9.**
*Input parameters.*

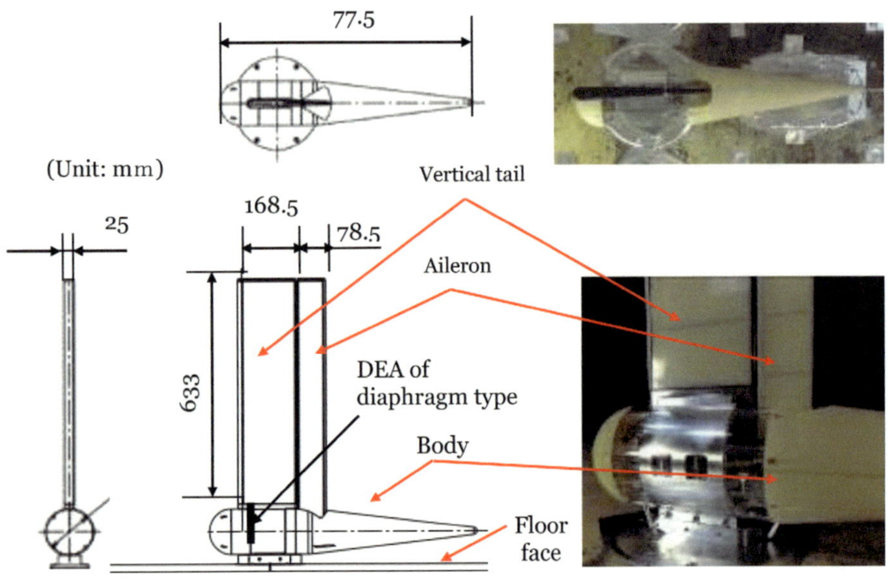


**Figure 10.**
*Body used in experiment.*

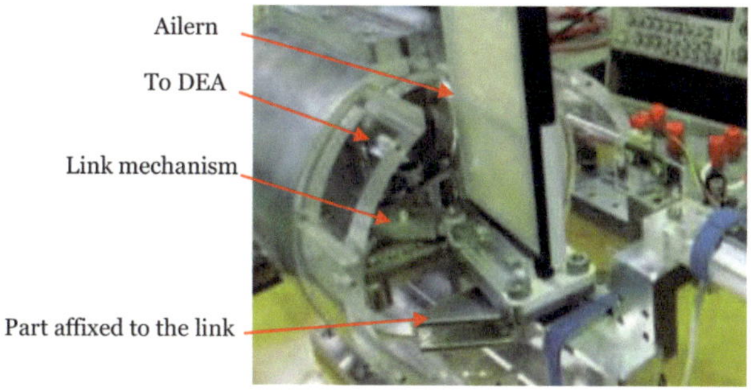


**Figure 11.**
*Link mechanism between the body and the measurement equipment.*

and the center is made of acrylic resin. For strength, the parts are reinforced with aluminum. The aileron has a hollow structure, but the total weight was about 5 kg. A DEA to control ailerons was installed inside the body. The DEA and aileron are connected by a link mechanism built into the main body (see **Figure 11**) [6], and when the DEA is displaced by 2 mm, the aileron moves 20 degrees. The DEA unit for aileron drive has a structure that adopts a diaphragm type with an outer dimension of φ100 mm.

As shown in **Figure 12**, a load measuring device other than the main body, a high-voltage power supply, a high-voltage switch, etc. were installed outside the wind tunnel and connected to the main body with a cable so as not to obstruct the air flow in the wind tunnel [6]. A video camera was installed on the ceiling outside

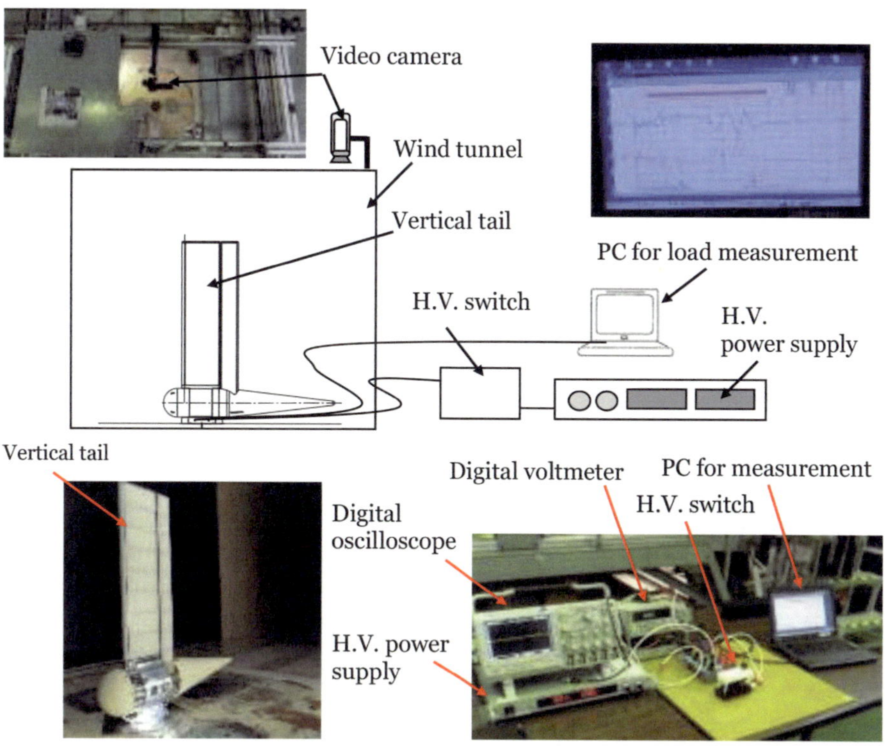


**Figure 12.**
*System diagram of the aileron drive experiment with a DEA.*

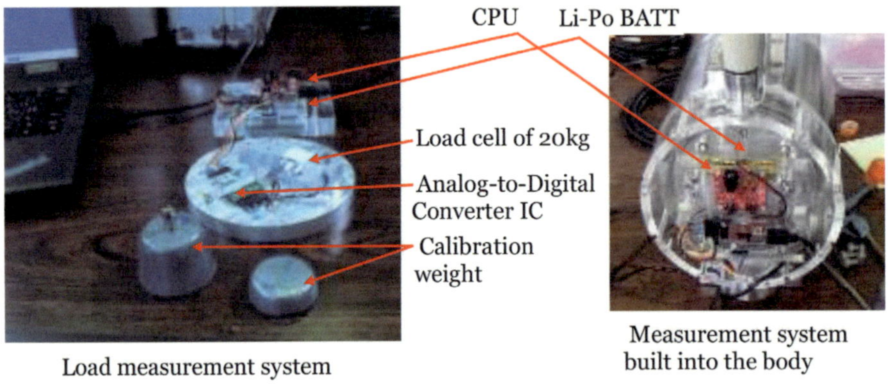


**Figure 13.**
*Load measurement system using a load cell.*

the wind tunnel to observe the state of the enclosure, and the images were taken from the observation window.

### 3.2 Preliminary experiment

First, in order to examine the specifications required to control the actuator used to steer the aileron, a load cell was used to measure the load applied to the actuator part at each steering angle. **Figure 13** shows the measurement system used for load measurement [6].

The load cell used for load measurement was mounted at the DEA mounting position of the main body. The maximum load applied to the load cell is 20 kg, and the analog data output from the load cell is converted to digital data by the 24-bit A/D converter IC (HX-711), and the CPU (ATmega328P). A lithium polymer battery (Li-Po) was used as the power source to minimize the effect of noise on the weak signal output from the load cell. In order to shorten the connection distance, the circuit to the CPU was attached to the fuselage. A personal computer for operation and recording was installed outside the wind tunnel, and it was connected by serial communication.

A laser distance sensor (VL50L0X) was also installed to measure the displacement velocity (**Figure 14**) [6]. Since this laser distance sensor uses I2C for the interface, it is not easily affected by noise and stable measurement is possible. The control was performed by the same CPU (ATmega328P) as the load measurement, and the measurement data was transmitted to the measurement PC by serial communication. Laser distance sensors were installed in the front and rear to observe the movement of ailerons and the DE.

The load was measured by continuously changing the wind speed from 0 m/s to 40 m/s in 5 m/s increments and setting the steering angles to 0, 5, 10, 15, and 20 degrees. When the set wind speed was reached, measurements were taken at each wind speed for about 30 seconds, and the load applied to the actuator section was measured (see **Figure 15**) [6].

In this experiment, it was found that the DEA was loaded with 11.54 kg/f in an environment with a steering angle of 20 degrees and a wind speed of 40 m/s (see **Table 1**). In order to move the aileron through the link mechanism, a force of about 2.6 kgf is required even in the absence of wind. It was found that a force of about 14.14 kg/f is required to steer 20 degrees in an environment with a wind speed of 40 m/s. It was also confirmed that the wings were deformed from the joint with the airframe at a wind speed of 30 m/s or more. Since the experiment was repeated in

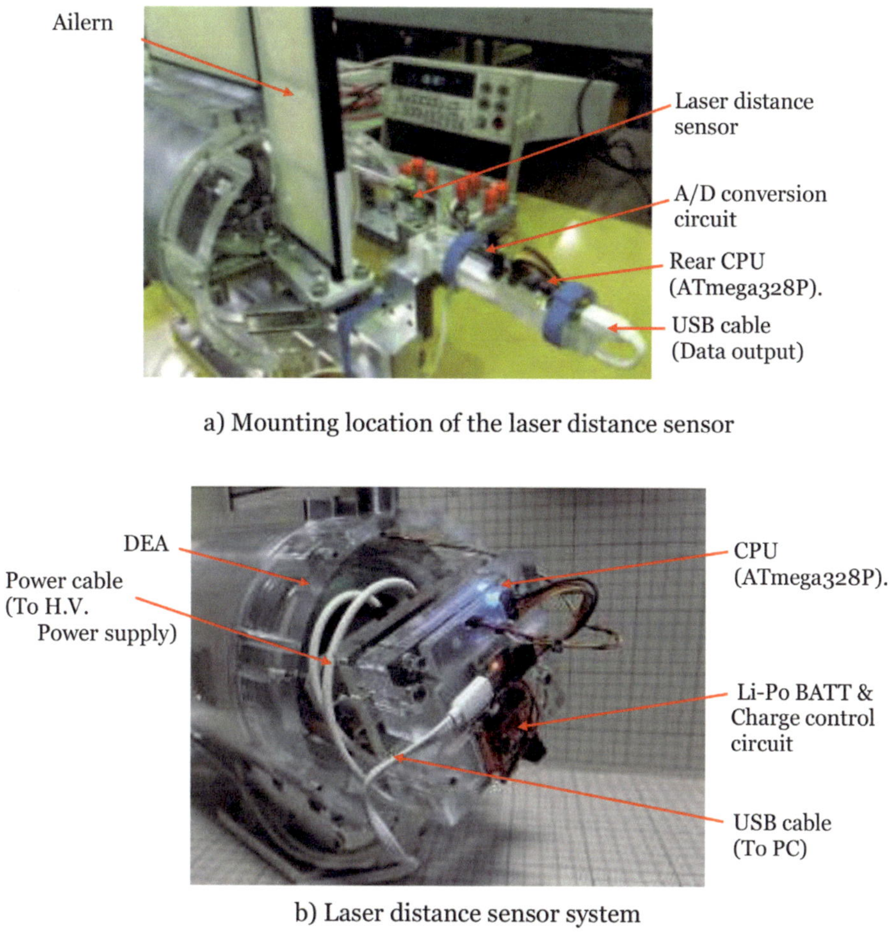


**Figure 14.**
*Laser distance sensor system installed inside the body. (a) Mounting location of the laser distance sensor; (b) Laser distance sensor system.*

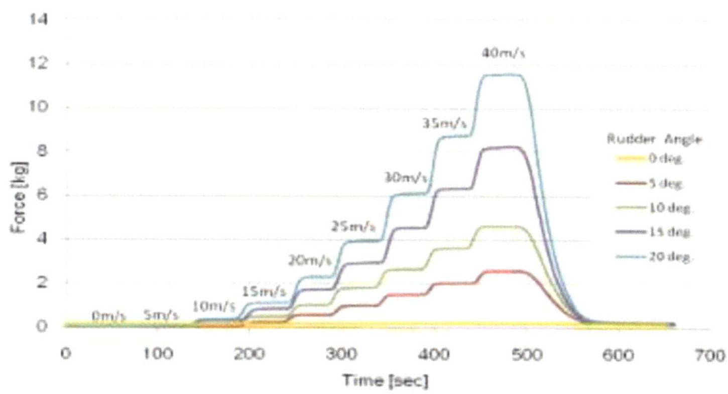


**Figure 15.**
*Load applied to the actuator section (by steering angle and wind speed).*

such a state and there was a risk of damage to the blade due to fatigue, the aileron drive experiment with the DEA was carried out at a wind speed of 0 m/s to 30 m/s. In order to steer 20 degrees in this environment, a DEA that can obtain a force of about 8.7 kg/f is required.

| | Wind speed (m/s) | | | | | | | |
|---|---|---|---|---|---|---|---|---|
| | 5 | 10 | 15 | 20 | 25 | 30 | 35 | 40 |
| Load applied to the actuator (kg / f) | 0.12 | 0.37 | 1.12 | 2.30 | 3.90 | 6.10 | 8.72 | 11.54 |

**Table 1.**
*Load applied to the actuator section at each wind speed when the rudder angle is 20 deg.*

### 3.3 Aileron drive experiment with a DEA

The load cell mounted for load measurement was replaced with a diaphragm type DEA, and an aileron drive experiment was conducted. **Figure 16** shows the state of the DEA mounting.

The DEA unit used had a donut shape with an outer diameter of φ100 mm, a DEA part with an outer diameter of φ80 mm, and a central part of φ50 mm (see **Figure 15a**). The DEA used a 3 M acrylic sheet (VHB4910) as the main elastomer and SWCNT (SG101) manufactured by Zeon Corporation as the main electrode material. This DEA unit had a displacement performance of 2.0 mm with an applied voltage of DC 3.2 kV under a load of 4.0 kg, and the drive time at this time was about 100 ms. The DEA was driven by a high voltage power supply and a high voltage switch located outside the wind tunnel. The high-voltage power supply installed outside the wind tunnel and the DEA installed in the enclosure are connected by a high-voltage cable with a length of about 6 m. However, due to the low current consumption of the DEA, the maximum voltage drop during driving is 100 V, which is within the range where there is no problem in driving the DEA. The DEA unit consists of four cartridges, and if one of the DEA cartridges fails, the remaining DEA cartridges can drive it.

In this experiment, the wind speed was changed from 0 m/s to 20 m/s every 5 m/s, and the change in wind speed at each wind speed was recorded with a video camera installed on the ceiling of the wind tunnel. The rudder angle was measured by analyzing the recorded video.

In the initial experimental plan, the angle of attack was planned to be changed from −10° to +10° in 5° increments, but the stress applied to the wing was greater than expected, so there was a risk of damage to the skeleton. In order to avoid such an outcome, this time, the angle of attack was set to 0° only, and the wind speed was changed from 0 to 20 m/s at 5 m/s intervals.

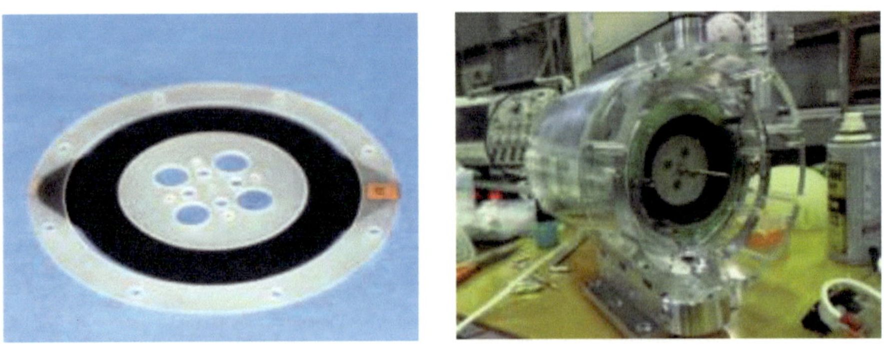

a) b)

**Figure 16.**
*DEA mounted in the body. (a) Diaphragm type DEA. (b) DEA built into the body.*

## 4. Results and discussion

**Table 2** shows the aileron angle at each wind speed. **Figure 17** shows the aileron displacement at each wind speed [6]. Up to a wind speed of 5 m/s, the aileron angle could be obtained up to 20 deg. However, the aileron angle gradually decreased from a wind speed of 10 m/s and reached 4 deg. at a wind speed of 15 m/s.

**Figure 18** shows the aileron drive speed when there is no wind, as measured in the lab. The rudder angle could be moved up to 20° at a speed of about 100 ms, and the same rudder angle and speed could be reproduced even if the drive control was repeated.

| | Wind speed (m/s) | | | | |
|---|---|---|---|---|---|
| | **0** | **5** | **10** | **15** | **20** |
| Aileron angle (deg.) | 20 | 20 | 9 | 4 | 0 |

**Table 2.**
*Aileron angle at each wind speed.*

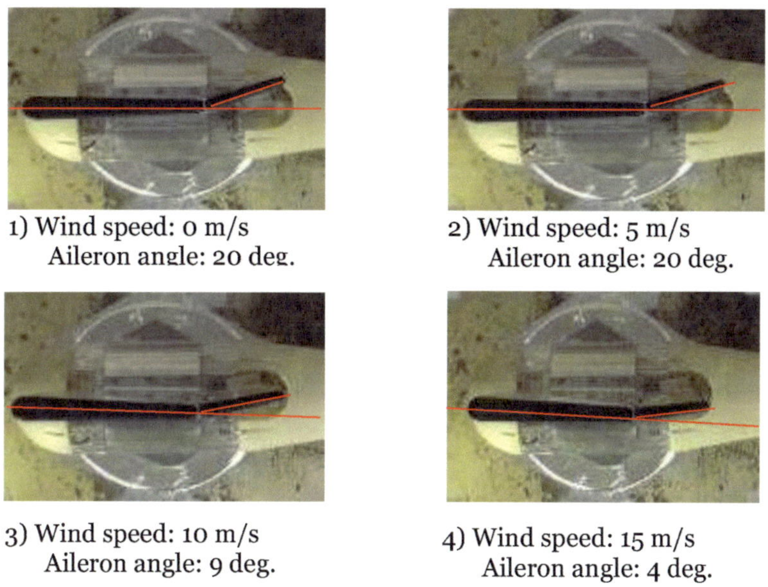


**Figure 17.**
*Aileron displacements at applied voltage DC3,200 V.*

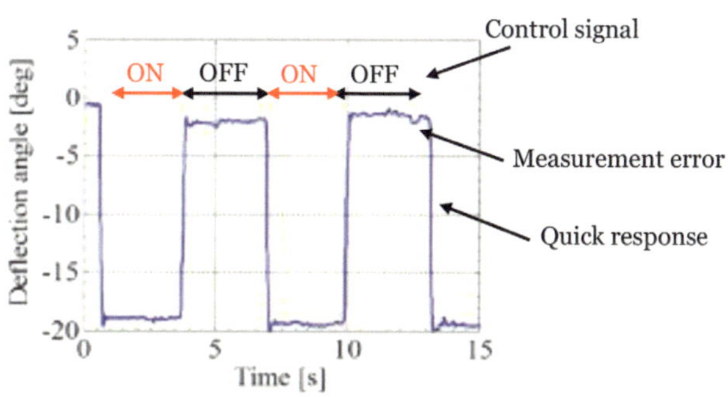


**Figure 18.**
*Aileron drive speed when there is no wind.*

**Figure 19** shows the operating speed of the aileron. The displacement speed was measured using a laser distance sensor mounted inside the body. At a wind speed of 5 m/s, it took 250 ms to reach the maximum displacement, but at a wind speed of 10 m/s, it took 300 ms. As described above, since the angle change of the aileron can be easily replaced with the voltage change, the feedback control can be easily performed by changing the voltage applied to the DEA using the voltage change.

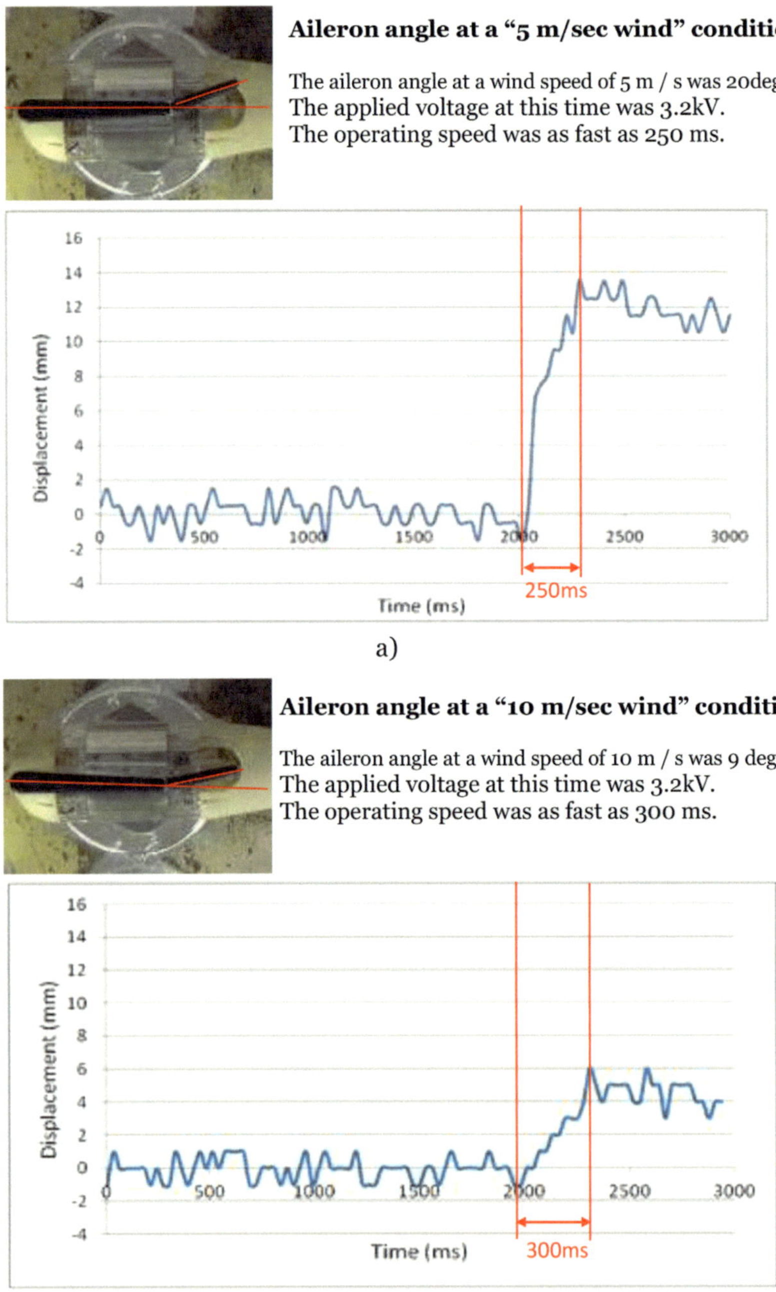


**Figure 19.**
*Aileron's operating speed measured by the laser distance sensor. (a) Wind speed of 5 m/s; (b) Wind speed of 10 m/s.*

The structural model used this time required a force of 2.6 kg for steering even at a wind speed of 0 m/s. This was a huge loss. It is probable that the maximum aileron angle could not be obtained due to insufficient driving force at a wind speed of 10 m/s. Most of this loss was due to the link mechanism. By increasing the efficiency of the link mechanism, we were able to obtain a maximum displacement of up to 15 m/s even with the same DEA. In the next experiment, we will investigate how much the mechanism can reduce the power consumed by developing a new drive that drives ailerons directly, enabling a simple and practical steering system.

To send a Martian plane to Mars, we need to dramatically reduce the weight of our compact and powerful motors. In addition, a powerful, efficient and responsive motor is essential for long-term flight of the spacecraft on the surface of Mars. Also, the surface temperature of Mars is very low and dust is present. Therefore, the required level of efficiency and responsiveness is very high. In this paper, based on the data obtained in this experiment, we attempted to compare the current level of a DEA with existing motors for these requirements.

First, we will explain the performance of the DE developed for this experiment:

The total weight of the DEA used is 52.8 g, of which 51.8 g is the weight of structures, etc., and the weight of the DEA itself is as small as 0.98 g. This DEA can lift a 4 kgf weight by 2 mm with an applied voltage of 3.3 kV. In order to increase this operating speed, the DE has been strengthened, and the total weight is 0.98 g, which is 98 ms.

Next is a comparison between the DE and existing motors:

From the above data, the power of the DE linear actuator is 0.0074 W per gram. As shown above, the weight including the DE actuator and its associated structure was approximately 53 g. If a similar linear actuator is configured using an existing DC motor of similar weight, the output of the linear actuator is 0.0015 (W). The weight including the DC motor and linear gear is about 95 g. Therefore, the DEA has a working speed per gram that is 4.9 times faster than a linear actuator that uses an existing DC motor. However, in the case of a linear actuator that uses a DC motor, a displacement of 1 mm takes about 200 milliseconds, so the difference in drive time is 9.9 times. In this experiment, we created a DE actuator that can lift a weight of 4 kgf using SWCNT (ZEONANO®-SG101) from Zeon Corporation. However, using high-crystal SWCNTs (extracted in the laboratory of Zeon Corporation under the guidance of Chiba et al.) gives about 1.32 times better results [35]. An SEM photograph of high-crystal CNTs is shown in **Figure 20**. It is estimated that DE motors using this CNT are about 13.1 times better than existing

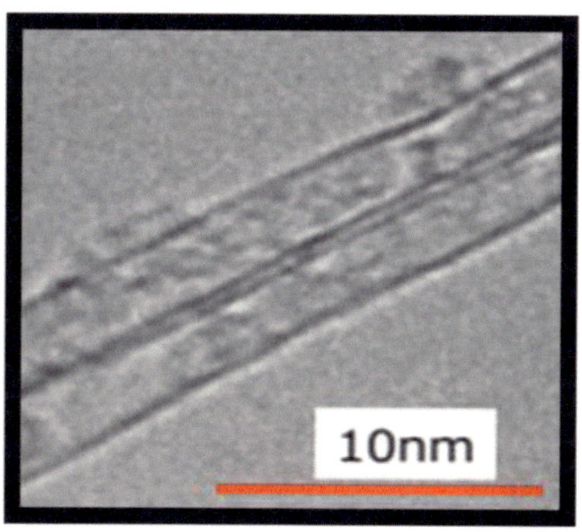


**Figure 20.**
*SEM photograph of high-crystal CNT.*

motors. When using metal CNTs, it is estimated to be twice as high as SWCNTs with high crystallinity. It is expected to be about 27.5 times that of existing motors.

In order to explain in more detail the good response of the DE obtained in this research, we compared it with the existing servo motor (which shows better performance than the model airplane used for radio control). The reason we chose the servo motor is that it can be controlled more accurately. The specifications of the servo motor that can obtain the same level of output as the DEA unit used this time are "servo motor (GWS): weight 41 g, torque 4.1 kg/cm, running speed: 270 ms /60 degrees". In contrast, the DEA used this time weighs 36 g (when four cartridges are built in), is about 13% lighter than the servo motor, and has a drive speed of 98 ms/2 mm, so it can be driven at higher speeds.

Based on the data obtained, the power consumption of the DE will be explained as follows. The power consumption during driving was measured with a voltage/current monitor of a high-voltage power supply installed outdoors. The wind speed was 5 m/s, the applied voltage was 3.2 kV, and the power consumption was 0.29 W. The current at this time was as small as 0.09 mA, and there was almost no voltage drop or heat generation due to the wiring cable. As mentioned above, one of the features of the DEA is that it consumes less current and can contribute to the weight reduction of the wiring cable, that is, the weight reduction of the main body. The power consumption during driving was measured with a voltage/current monitor of a high-voltage power supply installed outdoors. The wind speed was 10 m/s, the applied voltage was 3.2 kV, and the power consumption was 0.29 W. The current at this time was as small as 0.09 mA, and there was almost no voltage drop or heat generation due to the wiring cable. One of the features of the DEA is that it consumes less current. It can also contribute to the reduction of wiring cables and reduce the weight of the aircraft.

Considering the manufacturing cost of a DE, the weight of the DE including reinforcement is 0.96 g, which is cheaper and lighter than the price of a general existing motor with the same output. The SWCNT used as the electrode material for the DE in this experiment has started being mass produced at ZEON, so it costs about 1,000 yen ($ 9.6) per gram. Also, since the amount used is about 0.1 g, it is 100 yen (9 cents). The 3 M acrylic used for the elastomer is 20 yen per gram, so even if 1 g is used, it costs 120 yen ($ 1.15), which is cheap enough. Also, as mentioned above, the DEA itself is sufficiently lightweight and compact, which is a great advantage when mounted on a rocket.

When the DE is actually transported to Mars, it passes through outer space, so the effects of cosmic rays cannot be ignored. Next year, we plan to conduct a DE exposure test at the International Space Station (ISS: see **Figure 7**) and observe its effects.

Finally, we will explain the further improvement of the DE. As shown in **Figure 4**, we succeeded in launching a weight of 8 kg with a DE of 0.15 g. Using this, it will probably be able to move smoothly even in the wind with a higher speed. In this experiment, SWCNT (SG101) was also used, but it has been found also that the use of highly crystallized SWCNTs further improves drive speed and output. Even if the above link mechanism is not improved significantly, wind experiments of 25 m/s or more can be performed. In addition, new acrylics are currently being synthesized by Chiba et al. These acrylics can be used at −40°C to 150°C and may be able to handle even the harsh temperatures found on Mars. Also, due to the sufficient withstand voltage of the film, the control unit of the Mars probe will be developed mainly using this elastomer.

## 5. Results and discussion

From the above experimental results, it is suitable for controlling the Mars probe because it can output a large amount of DE even with a small number of DEs and has a high operating speed. In addition, DEs are manufactured at low manufacturing cost and can withstand −40°C.

Thus, for the first time in the world, we aim to fly in the atmosphere of Mars. By doing so, we would like to obtain our own Mars observation data such as high-resolution residual magnetic field data and atmospheric data. By establishing this technology, it will lead to the flight of other celestial bodies (Titan, Venus, etc.) that have an atmosphere.

The function of detecting radiation on Mars (particle type, energy range, dose range, etc.) and the function of detecting underground features such as caves and water volume, which are important considerations for Martian colonies, have not been considered in the paper. At this moment, the payload of the airplane is small and we will not able to consider them. As mentioned in the background of DEs, however, the output of the DE has come to lift 8 kg against the weight of the 0.15 g DE. In the near future, we would like to increase the output of the DE, so that a larger payload can be realized.

DEs are also suitable for controlling the solar panels, antenna control, and drive output control of planetary exploration spacecraft.

What we need to know in the future is how well DEs can withstand cosmic rays. As early as next year, we plan to experiment with how a DE behaves in space on the International Space Station.

Recently, more and more research has been aimed at exploring the possibility of applying DEs to frequently used items such as spacesuits, power suits, and robots, and we hope that the results of this research will be valuable.

## Acknowledgements

We would like to thank to Mr. Hiroki Ura and Mr. Takashi Yajima of the Aerodynamics Research Unit, Japan Aerospace Exploration Agency (JAXA)'s Ministry of Education, Culture, Science and Technology's Advanced Research Platform Operation Promotion Project "Wind and fluid Engineering Platform" for their enthusiastic support in getting this important data using their wind tunnel.

We also thank Aisin AW Co., LTD. for its financial support and manpower support during measurement in the wind tunnel.

## Author details

Seiki Chiba[1*] and Mikio Waki[2]

1 Chiba Science Institute, Yagumo, Meguro Ward, Tokyo, Japan

2 Wits Inc., Oshiage, Sakura, Tochigi, Japan

*Address all correspondence to: epam@hyperdrive-web.com

## References

[1] Nagai H, Oyama A: Mars Airplane, W.G. (2013) Mission Scenario of Mars Exploration by Airplane. Proceedings of the 2013 Asia-Pacific International Symposium on Aerospace Technology. 2013; 8:1-3.

[2] Fujita K, Luong R, Nagai H, Asai K: Conceptual Design of Mars Airplane, Trans. JSASS Aerospace Tech. Japan. 2012; 10: 28p.

[3] Fujita K, Nagai H, Robustness Analysis on Aerial Deployment Motion of a Mars Airplane using Multibody Dynamics Simulation: Effects of Wing-Unfolding Torque and Timing, The Aeronautical Journal. 2017; 121,1238: 449-468, DOI: 10.1017/aer.2016.123.

[4] Fujita K, Chiba S et al, Feasibility of artificial muscle for mars airplane. Aeron Aero Open Access J: 2018; 111: p213, DOI:10.15406/aaoaj.2018.02.00052.

[5] Fujita K, Oyama O, Kubo D, Kanazaki M, Nagai H, Wind Tunnel Test for Videogrammetric Deformation Measurement of UAV for Mars Airplane Balloon Experiment-1 (MABE-1), Journal of Flow Control, Measurement & Visualization. 2019; 7:87-100, DOI: 10.4236/jfcmv.2019.72007

[6] Chiba S et al, The challenge of controlling a small Mars exploration plane, In Proc. of SPIE, (Smart Structures and Materials Symposium and its 22nd Electroactive Polymer Actuators and Devices (EAPAD) Conference); 2020: 1137506. DOI: 10.1117/12.2551042.

[7] Katchalsky K, Rapid swelling and deswelling of reversible gels of polymeric acid by ionization. Experimentia,1945; 5: 319-320.

[8] Steinberget IZ et al, Mechanochemical engines, Nature; 1966: 210: 568-571.

[9] Hamilen F et. al, Electrolytically active contractile polymer: Nature, 1965; 206:1149-1150.

[10] Agolin FI, Gay FP, Synthesis and properties of azoaroumatic polymers. Macromolecules 3: 349-351,1970.

[11] Osada Y, Sato Y, Mechanochemical energy conversion in a polymer membrane by thermo-reversible polymer-polymer interactions. Makromolekulare Chem 1975; 176: 2761-2764.

[12] Tanaka T, Collapse of gel and the criticalendpoint. Phys Rev Lett 40; 1978: 820-823.

[13] Pei Q et. al, Electrochemical application of the bending beam method. 1. Masstransport and volume changes in polypyrrole during redox. J Phys Chem 1992; 96:10507-10514.

[14] Okuzak H, Kunugi T, Adsorption – induced bending of polypyrrole, films and its application to chemomechanical rotor. J. Polymer Sci. Polymer Phys. 1996; 43:1747-1749.

[15] Hirai T, Actuator materials from polymer gels. Polymer gels responding to electric and magnetic field. J. Matter Sci. Soc. Japan. 1995; 32:59-63.

[16] Pelrine R and Chiba S, Review of Artificial Muscle Approaches, (Invited) Proc. Third International Symposium on Micromachine and Human Science, Nagoya, Japan. 1992 October: 1-9.

[17] Yu Y et. al, Directed bending of a polymer film by light. Nature. 2003; 425:145.

[18] Oguro K, Fujiwara N, Asaka K, Onishi K, Sewa S, Polymer electrolyte actuator with gold electrodes, Proceedings of the SPIE's 6th Annual International Symposium on Smart

Structures and Materials, SPIE. 1999; 3669: 64-71.

[19] Otero FT, J. M. Sansiñena, Soft and wet conducting polymers for artificial muscles, Advanced Materials. 1998;10,6: 491-494.

[20] Osada Y, Okuzaki H, Hori H, A polymer gel with electrically driven motility, Nature, 1992; 355: 242-244.

[21] Wani OM, Zeng Z, Wasylczyk P, A. Priimagi, Programming Photoresponse in Liquid Crystal Polymer Actuators with Laser Projector, Journal Recommendation service. 2017; DOI: org/10.1002/adom.201700949

[22] Lee J et al, Magnetic force enhancement in a linear actuator by air-gap magnetic field distribution optimization and design, Finite Elements in Analysis and Design. 2012; 58: 44-52.

[23] Chiba S, Dielectric Elastomers, In Soft actuators, Asaka K, Okuzaki H, editors, Soft actuator, Springer; 2014. p. 183-195: ISBM978-4-431-54766-2.

[24] Tomori H et al. Theoretical Comparison of Mckibben-Type Artificial Muscle and novel Straight-Fiber-Type Artificial Muscle, Int. J. Automation Technology 2011: 5(4), 544

[25] Shintake J, Rosset S, Schubert B, Mintchev S, Floreano D, Shea H, DEA for soft robotics: 1-gram picks up a 60-gram egg, Proc. SPIE 9430, Electroactive Polymer Actuators and Devices (EAPAD); 2015: 94301S, DOI: 10.1117/12.2084043

[26] Anderson L, Gisby T, McKay T, O'Brienl B, Calius E, Multi-functional Dielectric Elastomer Artificial Muscles for Soft and Smart Machines, J. Appl. Phys;2012: 112,4: 041101.

[27] Kovacs GM, Manufacturing polymer transducers: opportunities and challenges, Proc. of SPIE; 2018:10594-10597.

[28] Pei Q, Dielectric Elastomers past, present and potential future, Proc. of SPIE; 2018:10594-10594.

[29] Pelrine R, Kornbluh R, Chiba S et al., High-field defomation of elasomeric dielectrics for actuators, Proc. 6th SPIE Symposium on Smart Structure and Materials; 1999: 3669, 149-161.

[30] Chiba S et al, Recent Progress on Soft Transducers for sensor Networks, In Technologies and Eco-innovation towards Sustainability II, Hu A, editor, Springer Nature; 2019: p285-298, DOI: org/10.1007/978-981-13-1196-3_23.

[31] Chiba S et al, Elastomer Transducers, Advances in Science and Technology, Trans Tech Publication, Switzerland; 2016: 97, 61-74. ISSN: 1662-0356, Doi: 10.4028/www scienctific.net/AST.97.61

[32] Chiba S et al, Challenge of creating high performance dielectric elastomers, Proc. of SPIE 2021 (Smart Structures and Materials Symposium and its 23rd Electroactive Polymer Actuators and Devices (EAPAD) Conference); 2021:1157-62

[33] Waki M, Chiba S, Elastomer Transducers, Soft actuators, Chapter 33, p 447-460, Springer, 2014, ISBM978-4-431-54766-2.

[34] Chiba S et al, Dielectric elastomer using CNT as an electrode, Proc. of SPIE 2020 (Smart Structures and Materials Symposium and its 22nd Electroactive Polymer Actuators and Devices (EAPAD) Conference); 2020:113751C–2. DOI:10.1117/12.2548512

Chapter 8

# Martian Moons and Space Transportation Using Chemical and Electric Propulsion Options

*Bryan Palaszewski*


## Abstract

Using chemical and nuclear electric propulsion for the exploration of the Martian moons will be investigated. Both oxygen/hydrogen chemical propulsion and nuclear electric propulsion with 500 kilowatt electric (kWe) to 10 megawatt electric (MWe) reactors will be assessed. The initial masses, propellant masses, and trip times for a variety of space vehicle payload masses will be compared. For high energy orbital transfer, the nuclear electric propulsion vehicles required a small fraction of the propellant mass over oxygen/hydrogen orbital transfer vehicles (OTVs). The moons, Phobos and Deimos, may hold resources for refueling future space vehicles. In-situ resource utilization (ISRU) can be a powerful method of reducing Earth dependence on space vehicle propellants, liquid water, and breathing gases. Historical studies have identified the potential of water in carbonaceous chondrites on the moons. The moon-derived propellants OTVs that move payloads between the moons and to other important operational Mars orbits. Also, the propellants have been suggested to support reusable Mars landers. To extract the water, the mined mass, its volume and the mining time were estimated. The water mass fraction may be as low as $2x10^{-4}$. Very large masses were needed to be extracted for up to 100 MT of water.


**Keywords:** Phobos, Deimos, Mars, electric propulsion, chemical propulsion, in-situ resource utilization, orbital transfer

## 1. Introduction

The Martian moons may be an important part of future Mars local planetary exploration and exploitation [1–11]. The moons have orbits that are relatively close to Mars, making them potential spacecraft berthing stations. Both oxygen/hydrogen (O2/H2) chemical propulsion and nuclear electric propulsion (NEP) orbital transfer vehicles were assessed. Their initial mass, propellant mass, and trip times were computed for several orbital transfer missions and orbital locations.

The moons may be sites of future in-situ resource utilization (ISRU), where metals and water may be wrested. The mining options for Phobos were assessed, showing the potential availability of water. A range of water mass fractions was investigated, and the potential masses of water were computed. The required mining time was also assessed.



IntechOpen

## 2. Mission design and options

### 2.1 Orbital transfer delta-V

Phobos and Deimos exploration and exploitation methods have been studied for many decades: landers, flybys, etc. [6–11]. While landers have been assessed in the past, this chapter will focus on the orbital transfer delta-V requirements and orbital transfer vehicle designs that would allow the 2 moons' exploration and exploitation.

The orbital missions are controlled by the delta-V or change of velocity needed for the orbit transfers. Both high-thrust missions and low-thrust missions were assessed. The high-thrust delta-V values were computed with a standard Hohmann transfer Equations [12]. The values for the low-thrust delta-V were calculated using the Edelbaum equation [13]. The nominal semi major axes for Phobos and Deimos are 9,378 and 22,459 km [2].

In all cases, the delta-V values are for round trip missions. There are 5 trips that were assessed: Phobos to Deimos, Phobos to areosynchronous Mars orbit (AMO), Phobos to 100,000 km altitude, Deimos to AMO, and Deimos to 100,000 km altitude.

**Figure 1** and **Table 1** provide the round trip delta-V for Phobos and Deimos missions. Both high thrust and low thrust delta-V values are presented. Due to the typical gravity losses with high thrust propulsion systems, a 20% delta-V increase

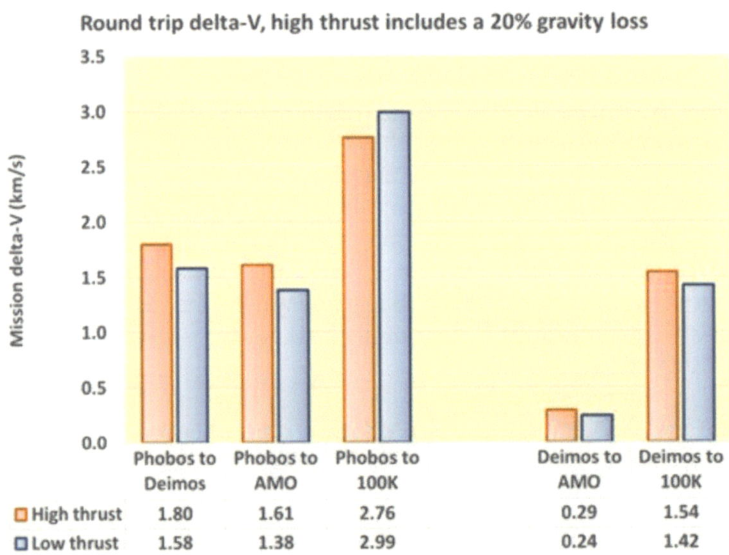


**Figure 1.**
*OTV delta-V, Phobos and Deimos orbital transfer missions.*

| Mission option | High thrust | Low thrust |
|---|---|---|
| Phobos to Deimos | 1.80 | 1.58 |
| Phobos to areocentric Mars orbit (AMO) | 1.61 | 1.38 |
| Phobos to 100,000 km altitude | 2.76 | 2.99 |
| Deimos to areocentric Mars orbit (AMO) | 0.29 | 0.24 |
| Deimos to 100,000 km | 1.54 | 1.42 |

**Table 1.**
*Orbital transfer mission options (for the high thrust options, the delta-V is increased by 20%).*

was added; no added losses were imposed on the low thrust systems. In **Figure 1**, the Phobos to 100,000 km low thrust delta-V was 2.99 km/s. The Phobos to Deimos low thrust delta-V was 1.58 km/s. At Deimos, the highest round trip delta-V is for the Deimos to AMO transfer was 0.24 km/s. The round trip low-thrust transfer to 100,000 km required only 1.42 km/s.

## 3. Propulsion options

High thrust chemical propulsion, using oxygen/hydrogen rocket engines is a natural choice [14]. If indeed water were available on the Martian moons, it would make sense to capitalize on that water resource, and finally producing rocket-purity oxygen and hydrogen.

Electric propulsion systems with either ion or Hall thrusters are potential options. Xenon or other inert gases are the typical choice for such thrusters. Using hydrogen as an electric propulsion propellant with a pulsed inductive thruster (PIT) has also been proposed.

### 3.1 Advanced propulsion options

Several advanced propulsion options for Martian moon transportation, exploration, and industrialization were investigated. Chemical propulsion and nuclear electric propulsion (NEP) with a range of power levels for Martian orbital transfer vehicles (OTVs) were assessed. Design parameters, vehicle mass scaling equations, and summaries of these analyses are presented; Mass scaling equations were developed for the O2/H2 chemical propulsion and the nuclear electric propulsion systems [14].

#### *3.1.1 Chemical propulsion OTV sizing*

In sizing the chemical propulsion OTVs, a vehicle mass scaling equation is used [14]:

$$m(dry, stage) = m(dry, coefficient)\, x\, m(p) + a(fixed) \qquad (1)$$

where

m(dry, stage) = the stage dry mass, including residual propellant (kg).

m(dry, coefficient) = the B mass coefficient (kg of tank mass/kg of usable propellant.

mass).

m(p) = usable propellant mass (kg).

a(fixed) = chemical OTV fixed mass (kg).

The chemical propulsion OTVs had a B coefficient of 0.4. The fixed mass was 500 kg. The fixed mass includes guidance systems, adapters and reaction control system masses. The Martian moon OTVs were single-stage vehicles.

#### *3.1.2 NEP OTV sizing*

The NEP OTV mass and trip time were estimated based on the power system and the propulsion system design [14]. The following dry mass scaling equation was used [12]:

$$m(dry.stage, NEP) = alpha\, x\, P + 0.05\, x\, m(p) + m(fixed) \qquad (2)$$

where
m(dry, stage, NEP) = NEP dry mass (kg).
alpha = NEP reactor specific mass (kg/kWe).
P = NEP power level (kWe).
0.05 = tankage mass coefficient (kg/kg m, p).
m(p) = NEP usable propellant mass (kg).
m(fixed) = NEP fixed mass (kg).

The OTV sizing was conducted for a wide range of power levels: 0.5 MWe to 30 MWe. Three nuclear reactor specific masses were used: 10, 20, and 40 kg/kWe (kilograms per kilowatt, electric) [15]. The OTV propulsion fixed mass, apart from and in addition to the reactor mass, was 20 MT, and the propellant tankage mass was 5% of the mass of the required propellant.

The specific impulse (Isp) and efficiency of the electric propulsion systems were 5,000 seconds with overall thruster-propulsion efficiencies of 50% for each design. These design points are typical of advanced designs of either magnetoplasmadynamic (MPD) or pulse inductive thrusters (PIT). While hydrogen is suggested for both propulsion system thrusters, the possibilities of the higher Isp option using inert gases (xenon, krypton, etc.) are also viable. The low thrust OTV delta-V value varied based on the destination of the Martian moon missions.

## 4. Mission effectiveness

### 4.1 Phobos and Deimos payload missions

A range of payload masses were included in the comparative orbital transfer cases: 1, 10 and 50 metric tons (MT). In general, the initial masses of the NEP OTVs are higher than the O2/H2 OTVs initial masses. However, the propellant masses of the NEP vehicles are generally significantly lower that most O2/H2 vehicle propellant masses. Thus, the propellant resupply masses for the NEP OTVs offer a substantial resupply mass benefit over chemical propulsion OTVs.

### 4.2 OTV mass comparisons

An initial comparison of the chemical and NEP option for the 1 MT payload cases is presented in **Figure 2**. Both the Phobos to Deimos and the Phobos to 100,000 km cases are shown. Overall, the initial masses of the NEP cases, for Phobos to Deimos and the Phobos to 100,000 km, are very similar; therefore, the larger Phobos to 100,000 km OTV NEP cases can perform both the Deimos and 100,000 km missions. The only NEP OTV designs that have a comparable propellant mass to the chemical propulsion OTV is the OTV with the 0.5 MWe power level.

The associated 1 MT payload OTV trip times for the chemical and NEP cases is presented in **Figure 3**. The Phobos to Deimos round trip time is for the 0.5 MWe case is 56.6 days. The 100,000 km round trip time is 108.9 days. The higher power levels provided a shorter trip time; however, the required propellant mass is higher than any chemical OTV propellant mass.

A summary of the initial masses of the chemical and NEP OTVs for the Phobos to Deimos and Phobos to 100,000 km is shown in **Figures 4** and **5**, respectively. The payload masses for both OTV mass estimates were 10 and 50 MT. The payload mass is carried on the full round trip mission In the Phobos to Deimos cases with a 10 MT payload mass, the benefit of the NEP system over the chemical OTV is best with NEP power levels of 0.5 to 1 MWe. For the 50 MT payload, the NEP OTV provides a very significant propellant mass benefit for power levels up to 10 MWe. For the

50 MT cases, the chemical OTV required about 31 MT of propellant, while the NEP OTV at a 10 MWe power level required only 9 MT.

The Phobos to 100,000 km orbital transfers are compared in **Figure 5**. Both the initial masses and propellant masses are shown. Comparisons are shown for 10 and 50 MT payload cases. The payload mass is carried on the full round trip mission. In general, the NEP OTV propellant mass savings over the chemical OTVs are very similar to the Phobos to Deimos cases. For the 50 MT cases, the chemical OTV required about 64 MT of propellant, while the NEP OTV at a 10 MWe power level required only 17 MT.

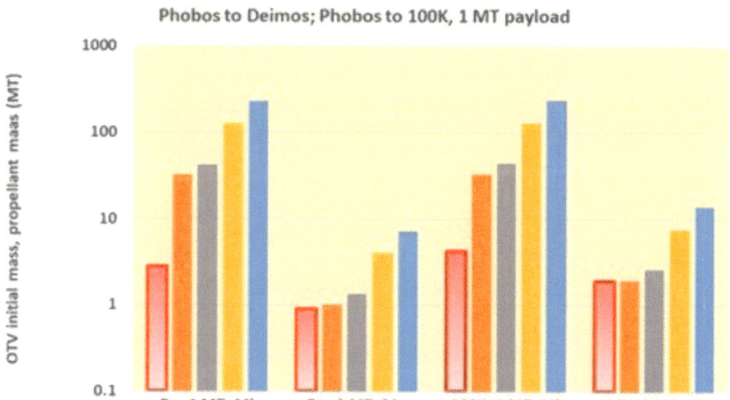


| | De, 1 MT, Mi | De, 1 MT, Mp | 100K, 1 MT, Mi | 100K, 1 MT, Mp |
|---|---|---|---|---|
| O2/H2 | 2.8 | 0.91 | 4.18 | 1.91 |
| NEP, 0.5 MWe | 32.1 | 1.02 | 33.1 | 1.96 |
| NEP, 1 MWe | 42.4 | 1.35 | 43.7 | 2.60 |
| NEP, 5 MWe | 125.2 | 4.00 | 129 | 7.63 |
| NEP, 10 MWe | 228.6 | 7.30 | 235.6 | 13.94 |

**Figure 2.**
*Initial mass and propellant resupply mass, Phobos to Deimos and Phobos to 100,000 km, round trip, 1 MT payload.*

| | De, 1 MT, t | 100K, 1 MT, t |
|---|---|---|
| O2/H2 | 0.74 | 4.71 |
| NEP, 0.5 MWe | 56.60 | 108.90 |
| NEP, 1 MWe | 37.40 | 72.00 |
| NEP, 5 MWe | 22.10 | 42.50 |
| NEP, 10 MWe | 20.20 | 38.80 |

**Figure 3.**
*Round trip time, Phobos to Deimos and Phobos to 100,000 km, round trip, 1 MT payload.*

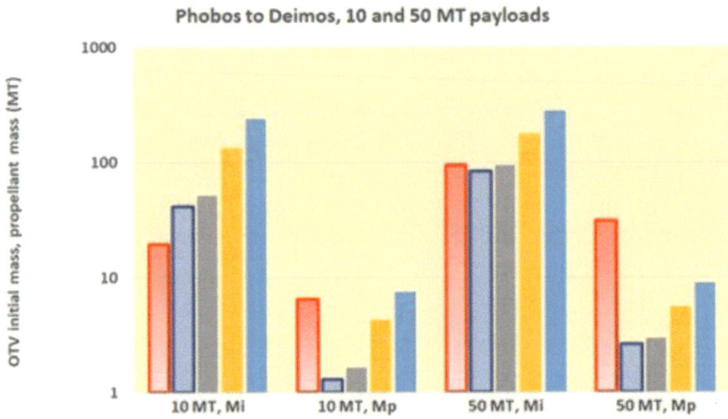


| | 10 MT, Mi | 10 MT, Mp | 50 MT, Mi | 50 MT, Mp |
|---|---|---|---|---|
| O2/H2 | 19.5 | 6.4 | 93.6 | 30.8 |
| NEP, 0.5 MWe | 41.4 | 1.31 | 82.8 | 2.6 |
| NEP, 1 MWe | 51.7 | 1.64 | 93.1 | 3.0 |
| NEP, 5 MWe | 134.5 | 4.26 | 175.9 | 5.6 |
| NEP, 10 MWe | 237.9 | 7.55 | 279.3 | 8.9 |

**Figure 4.**
*Initial mass and propellant resupply mass, Phobos to Deimos, round trip, 10 and 50 MT payload.*

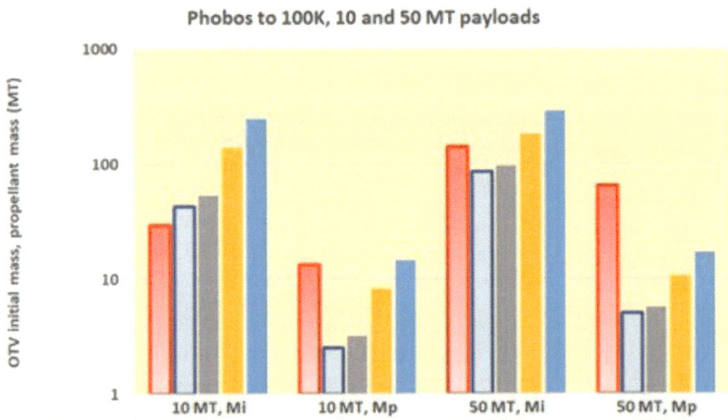


| | 10 MT, Mi | 10 MT, Mp | 50 MT, Mi | 50 MT, Mp |
|---|---|---|---|---|
| O2/H2 | 29.2 | 13.37 | 140.5 | 64.28 |
| NEP, 0.5 MWe | 42.7 | 2.52 | 85.3 | 5.05 |
| NEP, 1 MWe | 53.3 | 3.15 | 96.0 | 5.68 |
| NEP, 5 MWe | 138.6 | 8.20 | 181.3 | 10.72 |
| NEP, 10 MWe | 245.3 | 14.50 | 287.8 | 17.03 |

**Figure 5.**
*Initial mass and propellant resupply mass, Phobos to 100,000 km, round trip, 10 and 50 MT payloads.*

The trip time for the Phobos to Deimos with a 50 MT payload is shown in **Figure 6** for three reactor specific masses: 10, 20 and 40 kg/kWe. The NEP power levels of 0.5 to 10 MWe are of interest; once the power level reaches 10 MWe, the OTV has gained the greatest trip time benefits over the lowest power levels of 0.5 MWe. This example was provided to show the influence of reactor power level and specific mass on the OTV trip time.

For space science missions, the 1 MT payload cases can be important for several reasons. A small payload may be left in orbit or on the surface of one of the moons. The NEP OTV can then conduct radar experiments in concurrence with the orbiting or landed payload. Based of ground based meteorite analyses and spectroscopic measurements, Phobos and Deimos may have a surface of carbonaceous chondrites.

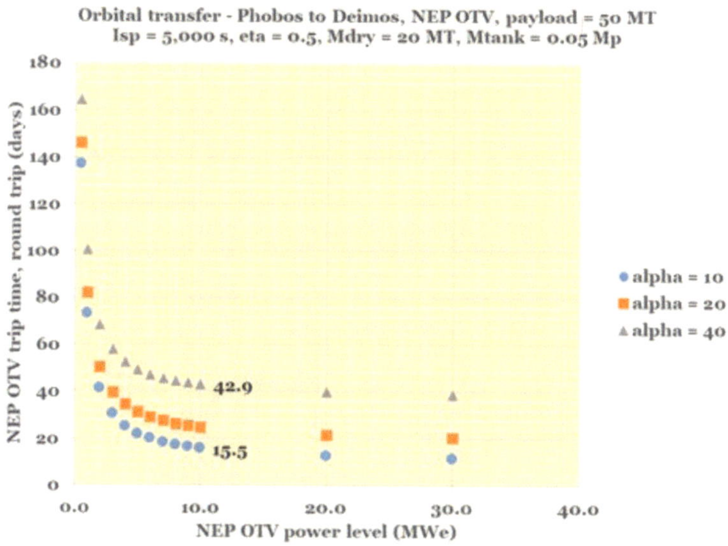


**Figure 6.**
*NEP round trip time versus power level, Phobos to Deimos, round trip, 50 MT payload.*

From this information, and from orbital gravity measurements, it is inferred that the moons may have a high porosity. The radar measurements can illuminate or knowledge about the moons' interior geological structures and the potential locations of frozen water reserves.

Detailed comparisons of the chemical and NEP OTV resupply propellant masses and specific trip times for the 10 MT payload cases are presented in **Figures 7–11**. The set of cases for the 50 MT payloads are presented in **Figures 12–16**. In general, the NEP trip times are many days, whereas the chemical OTV trip times are much shorter. The Phobos to 100,000 km orbit transfer required the largest mission delta-V and the largest OTVs; therefore, this OTV design can encompass all the suggested OTV missions.

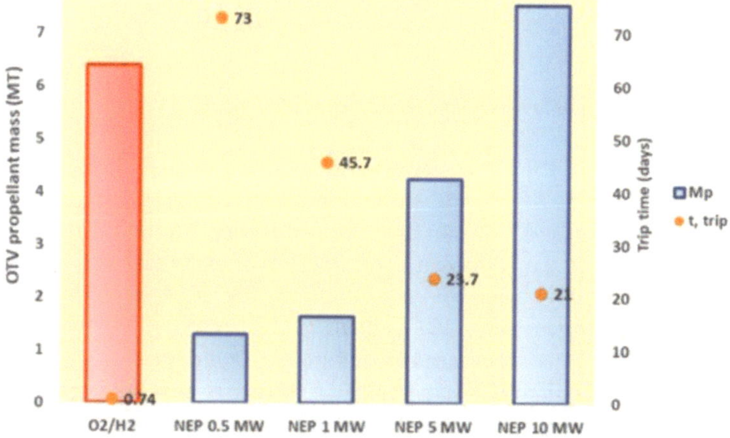


**Figure 7.**
*Propellant resupply mass, Phobos to Deimos, round trip, 10 MT payload.*

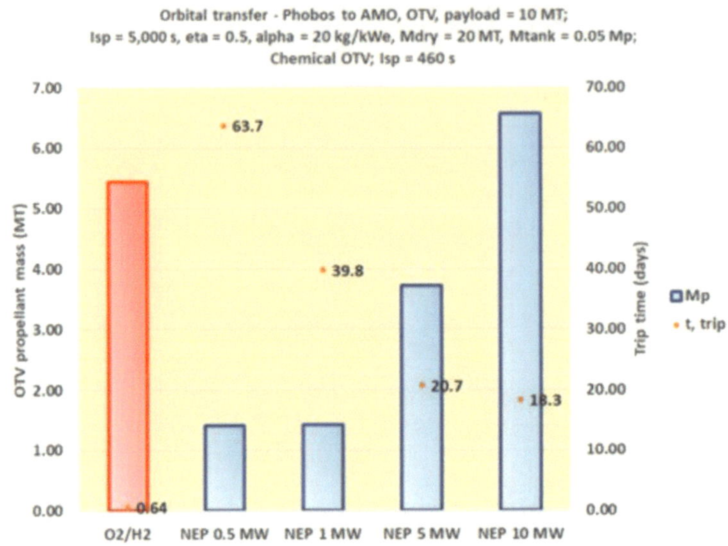


**Figure 8.**
*Propellant resupply mass, Phobos to AMO, round trip, 10 MT payload.*

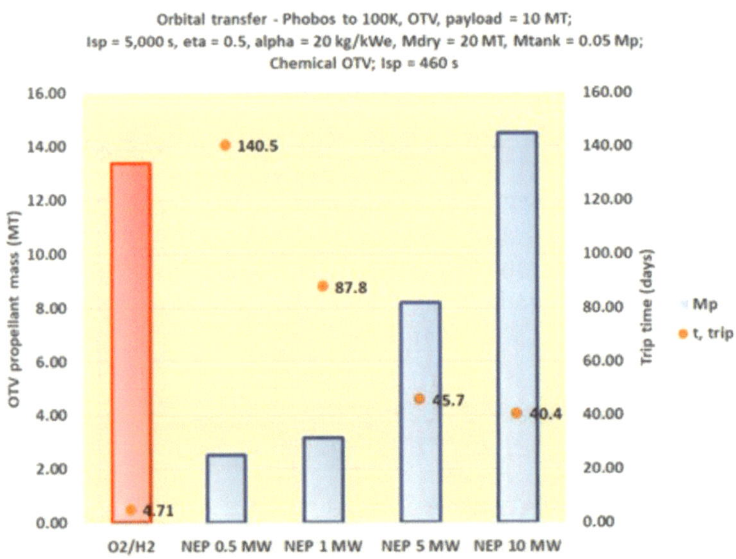


**Figure 9.**
*Propellant resupply mass, Phobos to 100,000 km, round trip, 10 MT payload.*

## 4.3 Martian moons and ISRU - water mining

Phobos has been studied in detail over many decades. Models of the moon have suggested that the surface may have a large fraction of carbonaceous chondrites. These chondrites may have a sizable water content. Preliminary estimates of the water mass fraction range from $1x10^{-5}$ to $1x10^{-1}$. The estimates were based on models and laboratory measurements of meteoritic chondrites.

If water in indeed available, it can be used to create resupply propellants for the Martian OTVs. In addition to the refueling of the NEP and chemical OTVs, Mars lander analyses (Mars Base Camp) [16] have shown a need for approximately 100 MT of water to create the required 78 MT of O2/H2 propellant. This 100 MT water mass was used as a guide for the ISRU analyses.

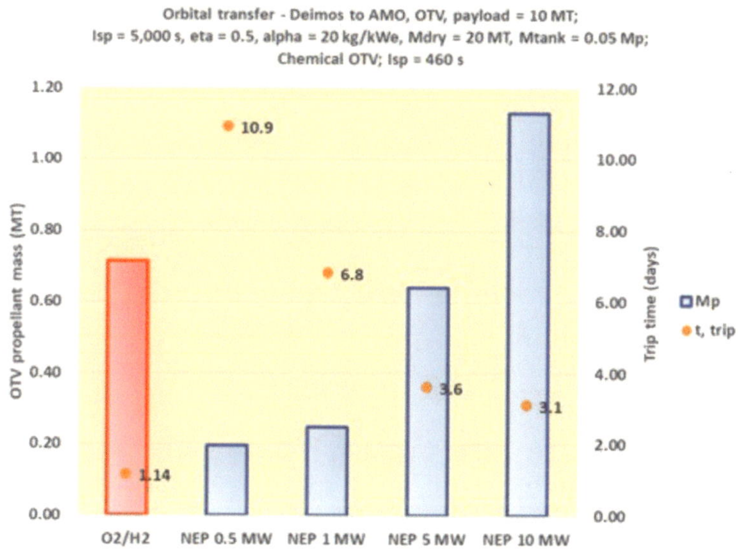


**Figure 10.**
*Propellant resupply mass, Deimos to AMO, round trip, 10 MT payload.*

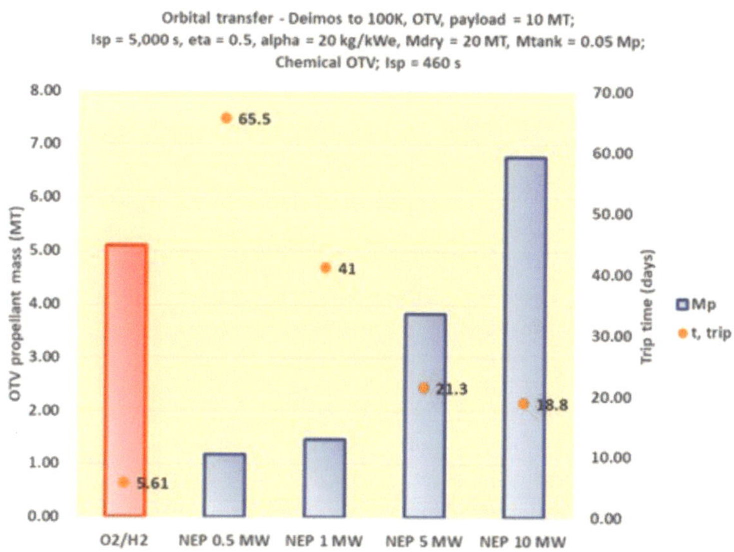


**Figure 11.**
*Propellant resupply mass, Deimos to 100,000 km, round trip, 10 MT payload.*

While water is an important commodity that may be wrested from the Martian moons, the mass of water for the chemical OTV propellant resupply can be very high. In future cases using pulsed inductive thrusters, hydrogen propellant can be used in NEP OTVs, and therefore benefit from such water reserves. With the high NEP Isp values, the propellant mass is much lower than that for chemical OTVs, significantly reducing the mining requirements.

### *4.3.1 Issues of water unavailability*

There has been much speculation regarding the water content of the Martian moons. Research programs have suggested that the moons agglomerated from the matter that formed Mars. The water content was estimated to be $2x10^{-4}$ (or 0.02

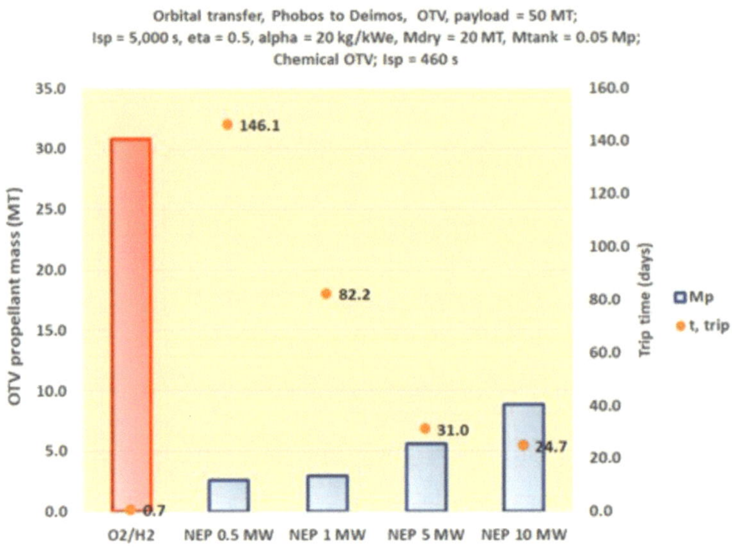


**Figure 12.**
*Propellant resupply mass, Phobos to Deimos, round trip, 50 MT payload.*

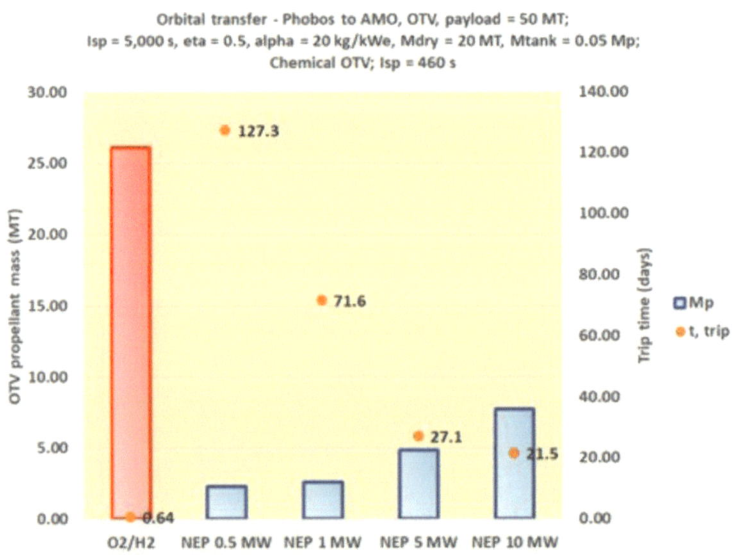


**Figure 13.**
*Propellant resupply mass, Phobos to AMO, round trip, 50 MT payload.*

weight%) [17]. Recent lunar water research has suggested widespread water on the Moon as being 1 to 4x$10^{-4}$ weight% [17]. Given the wide range of possible water mass fractions, analyses were conducted using a mass fraction of 1x$10^{-5}$ to 1x$10^{-1}$. **Figure 17** shows the water mass that may be available on Phobos. For simplicity, the radius of Phobos was assumed to be 9 km. The area mined is 10 x 10 meters and 1 meter deep. With the lowest mass fraction of 1x$10^{-5}$, the total water available would be approximately 18,000 MT; implying that approximately one hundred and eighty (180), 100 MT water loads can be extracted. For the mass fraction of $10^{-5}$, the area to be mined is 180th the moon's surface area: approximately 5.66 $km^2$.

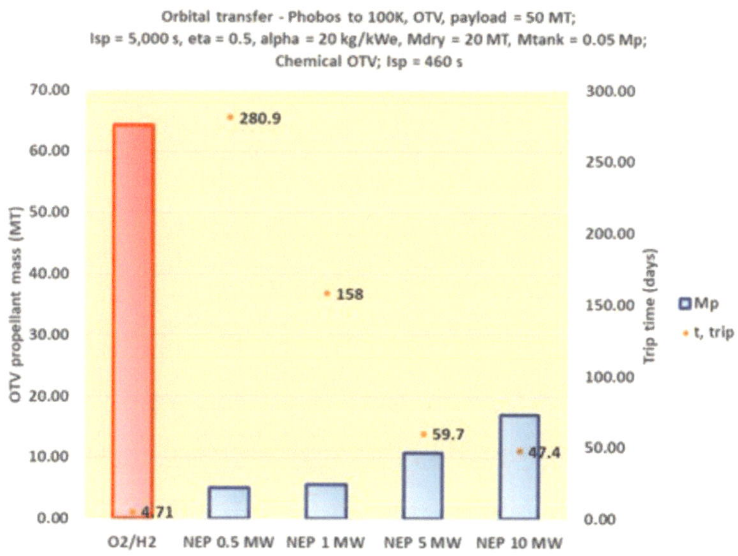


**Figure 14.**
*Propellant resupply mass, Phobos to 100,000 km, round trip, 50 MT payload.*

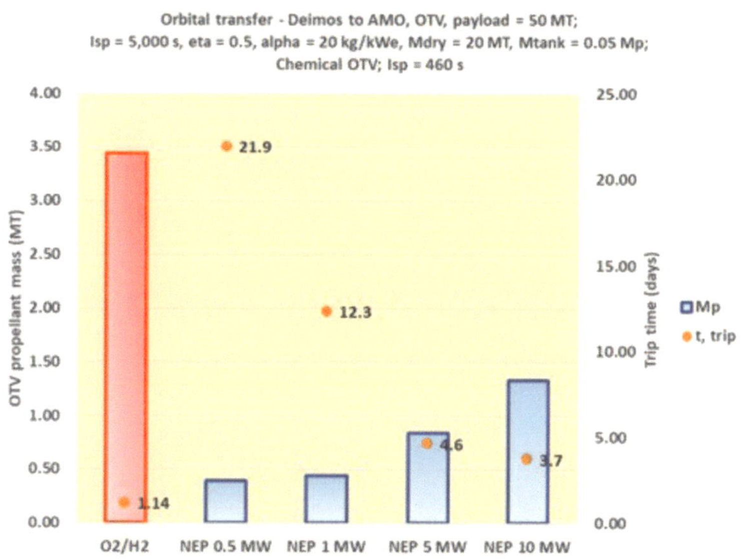


**Figure 15.**
*Propellant resupply mass, Deimos to AMO, round trip, 50 MT payload.*

The mined water mass is a very small fraction of the total regolith to be processed. The volume of the mined mass or radius of a proposed spherical mining container was computed and shown in **Figure 18**. For the mass fraction of $1x10^{-2}$, the capture tank radius would be 11 meters; for the $1x10^{-5}$ mass fraction, the radius would be 110 meters. Separation of the water from the total mined mass will be quite a challenge; the water and the regolith must be separated in the very low gravity field on the moons. The water and the final production propellant purity must be maintained to make the ISRU-based propulsion systems a success.

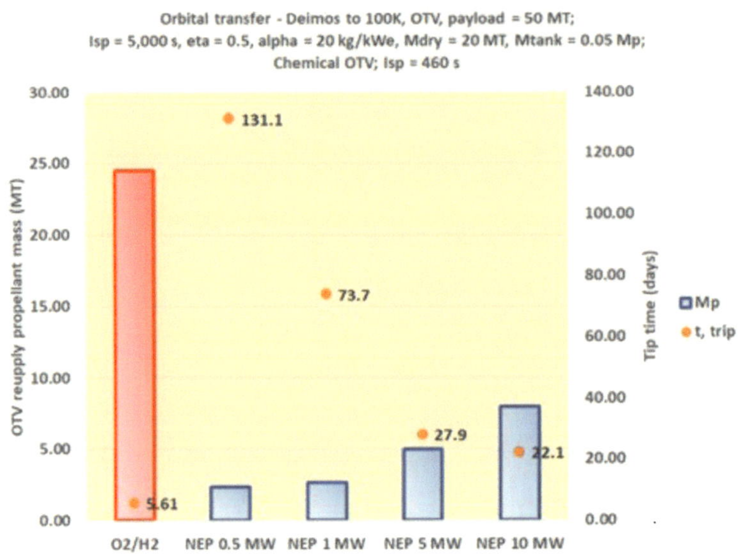


**Figure 16.**
*Propellant resupply mass, Deimos to 100,000 km, round trip, 50 MT payload.*

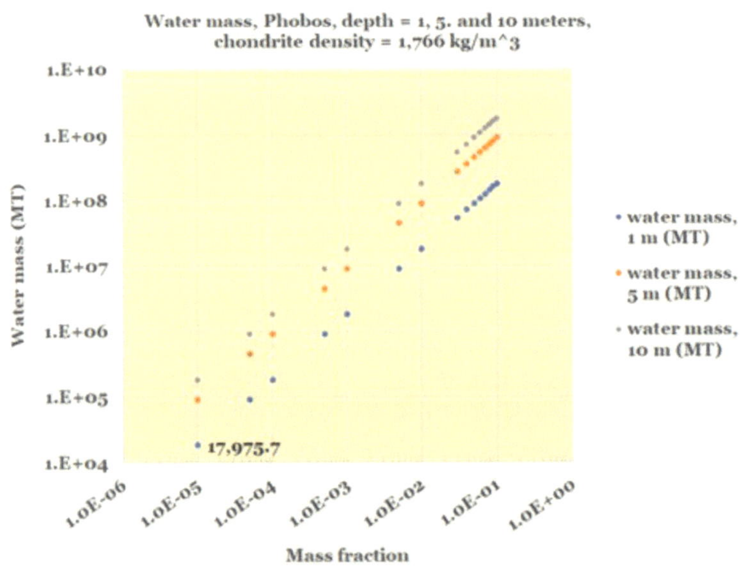


**Figure 17.**
*Water mass predictions, Phobos, water mass fraction: 1x10$^{-5}$ to 1x10$^{-1}$.*

The Phobos water mining time is shown in **Figure 19**; the figure shows the time needed to extract a wide range of water masses. If the mass fraction is 1x10$^{-2}$, the mining time is approximately 57 days to extract 100 MT. For the 1x10$^{-5}$ mass fraction, the mining time is 57,000 days. Thus, only the higher the mass fractions will be useful for large scale water production.

Once the water mass fraction is established, more effective planning and designing of the mining machines will be possible. One possibility is that the water may exist as ice deep inside Phobos [18]. Reference 17 notes that the ice location may be 10 to 100 meters below the surface. Extracting the water would therefore require a very sophisticated mining system, far more complicated than any surface mining

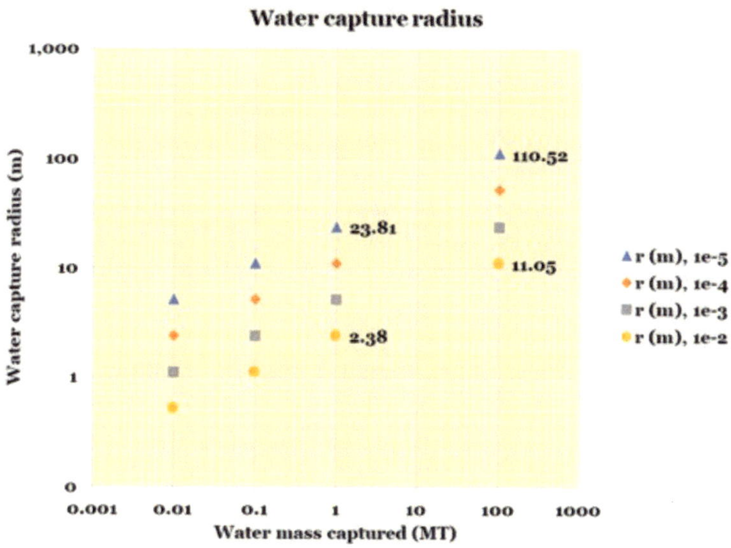


**Figure 18.**
*Water mining storage vessel radius, Phobos, water mass fraction: $1x10^{-5}$ to $1x10^{-2}$.*

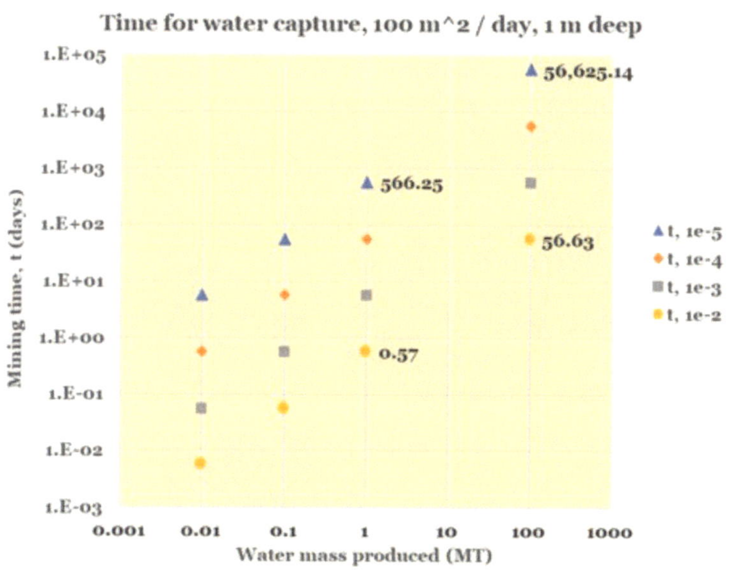


**Figure 19.**
*Water mining time, Phobos, water mass fraction: $1x10^{-5}$ to $1x10^{-2}$.*

system. If the moons' surfaces do not possess any water, then the metal and other raw materials would be the best Phobos ISRU products.

## 5. Conclusions

For exploration and exploitation of the Martian moons, both chemical propulsion and electric propulsion orbital transfer vehicles (OTVs) were assessed. For large payloads of 10 to 50 MT, the nuclear electric propulsion (NEP) OTVs require a small fraction of the chemical propulsion OTV propellant mass. If 10 MT payload masses can be manifested together, the 50 MT NEP OTV would be a more propellant efficient OTV option over a using five 10 MT NEP OTVs. The Phobos to 100,000 orbit transfer required the largest mission delta-V and the largest OTVs; therefore, this OTV design can encompass all the suggested OTV missions.

For small 1 MT payloads, chemical propulsion OTVs were more efficient than NEP OTVs. However, the NEP OTV can enable several special missions. Radar science missions and observations can be conducted; a small 1 MT payload might be left in orbit about a Martian moon and the NEP OTV can use a high power radar to transmit signals through the moon. The smaller payload would gather the reflected radar signal and allow more accurate determination of the moons internal structure.

Mining water on the Martian moons may be used for resupplying propellants to chemical and NEP OTVs. Mining systems on the surface may be able to provide the needed water to make hydrogen and oxygen. However, the water mass fraction on the surfaces Phobos and Deimos may be small; the estimates have a wide range from $1x10^{-5}$ to $1x10^{-1}$. Deep caches of water ice may also exist, but these deep caches may be difficult to mine. In the best estimates, 100's of MT of water may be wrested from Mars' moons, assisting in make Mars exploration truly Earth independent.

## Nomenclature

| | |
|---|---|
| a | Acceleration of gravity |
| AMO | Areosynchronous Mars orbit |
| delta-V | Velocity change |
| g | Gravity level (compared to Earth) |
| H2 | Hydrogen |
| Isp | Specific impulse |
| ISRU | In-situ resource utilization |
| m(p) | Propellant mass |
| m(pl) | Payload mass |
| NEP | Nuclear electric propulsion |
| O2/H2 | Oxygen/hydrogen |

## A. Appendix

### Phobos maps:

https://astrogeology.usgs.gov/search/map/Phobos/Viking/Phobos_Viking_Mosaic_40ppd_DLRcontrol

https://astrogeology.usgs.gov/search/map/Phobos/MarsExpress/SRC/Phobos_ME_SRC_Mosaic_Global_16ppd

https://www.sciencedirect.com/science/article/abs/pii/S0032063313001293

## Bibliography

P. Aftabi, "THE WATER TRACES AND STRUCTURAL LINEAMENTS ON MARTIAN MOONS," Geological Survey of Iran, LPI Contribution No. 1377.

D. T. Britt, D. Yeomans, K. Housen, G. Consolmagno, "Asteroid Density, Porosity, and Structure," Book, Asteroids III.

Bruck Syal, M., J. Rovny, J. M. Owen, and P. L. Miller (2016), "Excavating Stickney Crater at Phobos, Geophys. Res. Lett., 43, 10,595–10,601, doi:10.1002/2016GL070749., 2016

B. Carry, "Density of asteroids." European Space Astronomy Centre, ESA, Preprint submitted to PSS March 21, 2012, arXiv:1203.4336v1.

G.J. Consolmagno, D.T. Britt, R.J. Macke, "INVITED REVIEW - The significance of meteorite density and porosity," Chemie der Erde 68 (2008) pp. 1–29.

FRASER P. FANALE AND JAMES R. SALVAIL, "Evolution of the Water Regime of Phobos," ICARUS 88, pp. 380-395 (1990).

PROPERTIES OF PHOBOS FROM THE MARS GLOBAL SURVEYOR THERMAL EMISSION SPECTROMETER: EVIDENCE FOR WATER AND CARBONATE T. D. Glotch, C. S. Edwards, and D. S. Ebel, 46th Lunar and Planetary Science Conference (2015), abstract 2587.

David J. Lawrence , Patrick N. Peplowski , Andrew W. Beck, Morgan T. Burks. Nancy L. Chabot, Michael J. Cully, Richard C. Elphic, Carolyn M. Ernst , Samuel Fix, John O. Goldsten, Erin M. Hoffer, Hiroki Kusano, Scott L. Murchie , Brian C. Schratz, Tomohiro Usui, and Zachary W. Yokley, "Measuring the Elemental Composition of Phobos: The Mars-moon Exploration with GAmma rays and NEutrons (MEGANE) Investigation for the Martian Moons eXploration (MMX) Mission," AGU earth and Space Science, RESEARCH ARTICLE, 10.1029/2019EA000811

S.L. Murchie, A.A. Fraeman, R.E. Arvidson, A.S. Rivkin, R.V. Morris, "INTERNAL CHARACTERISTICS OF PHOBOS AND DEIMOS FROM SPECTRAL PROPERTIES AND DENSITY: RELATIONSHIP TO LANDFORMS AND COMPARISON WITH ASTEROIDS," 44th Lunar and Planetary Science Conference (2013), abstract 1604.

M. Nakajima and R. M. Canup," Origin of the Martian Moons and Their Water Abundances." Lunar and Planetary Science XLVIII, Abstract 2900, (2017).

Rosenblatt P., Hyodo R., Pignatale F., Trinh A., Charnoz S., Dunseath K.M., Dunseath-Terao M., & Genda H., "The formation of the Martian moons," The Final Manuscript to Oxford Science Encyclopedia, Oxford University Press, March 2020.

Alex Soumbatov-Gur. "Phobos, Deimos: Formation and Evolution." [Research Report] Karpov institute of physical chemistry. 2019. hal-02147461, https://hal.archives-ouvertes.fr/hal-02147461, Submitted on 4 Jun 2019.

Scheeres et al., "Heterogeneous mass distribution of the rubble-pile asteroid (101955) Bennu," Sci. Adv. 2020; 6 : eabc3350, 8 October 2020.

Megan Bruck Syal, Jared Rovny, J. Michael Owen, and Paul L. Miller, "Excavating Stickney Crater at Phobos," Geophysical Research Letters, 08 October 2016.

## Author details

Bryan Palaszewski
Engine Combustion Branch, NASA Glenn Research Center, Cleveland, OH, USA

*Address all correspondence to: bryan.a.palaszewski@nasa.gov

## References

[1] Sheehan, W., The Planet Mars: A History of Observation & Discovery, University of Arizona Press, 1996

[2] Mars fact sheet, National Space Science Data Center, nssdc.gsfc.nasa.gov/planetary/factsheet/marsfact.html

[3] Veverka, Joseph; Noland, Michael; Sagan, Carl; Pollack, James; Quam, Lynn; Tucker, Robert; Eross, Botand; Duxbury, Thomas; Green, William (1974). "A Mariner 9 Atlas of the Moons of Mars," Icarus 23 (2): pp. 206-289.

[4] Bernard Godard, Frank Budnik, Pablo Munoz, Trevor Morley, and Vishnu Janarthanan, "ORBIT DETERMINATION OF ROSETTA AROUND COMET 67P/CHURYUMOV-GERASIMENKO," 2015. https://issfd.org/2015/files/downloads/papers/124_Godard.pdf

[5] Godard B, Budnik F, Muñoz P, et al. Orbit determination of Rosetta around Comet 67P/Churyumov-Gerasimenko. In: Proceedings of 25th International Symposium on Space Flight Dynamics. Munich, 2015

[6] Anon., "First International Conference on the Exploration of Phobos and Deimos," LPI contribution number 1377, held at NASA Ames Research Center, November 5-7, 2007.

[7] Bosanac, N., et al., "Manned Sample Return Mission to Phobos: a Technology Demonstration for Human Exploration of Mars," 2014 IEEE Aerospace Conference, June 2014.

[8] Anon., "A Study of System Requirements for Phobos / Deimos Missions. Volume II, Phase I Results – Satellite Rendezvous and Landing Missions," NASA Contract NAS1-10873, Martin Marietta Corp., June 1972.

[9] Abercromby, A., et al., "Human Exploration of Phobos," 2015 IEEE Aerospace Conference, 7-14 March 2015.

[10] Pascal, Lee, et al., "First International Conference on the Exploration of Phobos and Deimos: The Science, Robotic Reconnaissance, nd Human Exploration of the Two Moons of Mars," Moffett Field, California, November 5-7, 2007. Lunar and Planetary Institute, LPI Contribution No. 1377

[11] Utashima, Masayoshi, "Design and Analysis of Phobos-Rendezvous Orbit," Systems Engineering Department, National Space Development Agency of Japan, Journal of the Japan Society for Aeronautical and Space Sciences, Vol. 44 No. 506, March 1996, Published by The Japan Society for Aeronautical and Space Sciences, National Space Development Agency of Japan, NASDA English Translation, December 1997,

[12] Roger R. Bate, Donald D. Mueller, Jerry E. White, "Fundamentals of Astrodynamics," Dover Books on Aeronautical Engineering, Jan 1971.

[13] Edelbaum, T. N., "Propulsion Requirements for Controllable Satellites," ARS Journal, Vol. 31, Aug. 1961, pp. 1079-1089.

[14] Palaszewski, B., "Atmospheric Mining of the Outer Solar System: Resource Processing Moon Base Propulsion, and Vehicle Design Issues," AIAA 2019-4031, August 2019.

[15] Mason, L., A Comparison of Energy Conversion Technologies for Space Nuclear Power Systems, AIAA 2018-4977, July 2018.

[16] Timothy Cichan, Sean O'Dell, Danielle Richey, Stephen A. Bailey, Adam Burch, "MARS BASE CAMP UPDATES AND NEW CONCEPTS," Lockheed Martin Corporation. IAC-17,A5,2,7,x40817, 68th International Astronautical Congress (IAC), Adelaide, Australia, 25-29 September 2017.

[17] Ngoc Truong, Pascal Lee, "Water on Phobos and Deimos: Implications of Water in Tektites for the Giant Impact Origin Hypothesis for the Moons of Mars." NASA's Third International Conference on the Exploration of Phobos & Deimos, 18-19 Jul 2016, NASA Ames Research Center, PhD2016-052. https://nesf2016.arc.nasa.gov/abstract/nesf2016-138.html

[18] FRASER P. FANALE AND JAMES R. SALVAIL, "Evolution of the Water Regime of Phobos," ICARUS 88, 380-395 (1990).

Section 5

# Asteroid Belt

Chapter 9

# Special Crater Types on Vesta and Ceres as Revealed by Dawn

*Katrin Krohn*


## Abstract

The exploration of two small planetary bodies by the Dawn mission revealed multifaced surfaces showing a diverse geology and surface features. Impact crater are the most distinctive features on these planetary bodies. The surfaces of asteroid Vesta and the dwarf planet Ceres reveal craters with an individual appearance as caused by different formation processes. Special topographic and subsurface conditions on both bodies have led to the development of special crater types. This chapter present the three most characteristic crater forms fund on both bodies. Asymmetric craters are found on both bodies, whereas ring-mold craters and floor-fractured craters are only visible on Ceres.


**Keywords:** asteroids, asymmetric craters, ring mold craters, floor-fractured craters, impact into ice

## 1. Introduction

The Dawn Mission was the first mission exploring two different planetary objects in the asteroid belt between Mars and Jupiter, the asteroid Vesta and the dwarf planet Ceres. Asteroid Vesta is the second most massive asteroid (2.59079 x $10^{20}$ kg) with a mean diameter of 525 km and a mean density of 3.456 ± 0.035 g/cm (e.g., [1, 2]). Vesta is believed to be a dry, differentiated proto-planet with an iron core of about 220 km in diameter, a mantle with diogenite compositions and an igneous crust [1, 3, 4]. Asteroid Vesta is a fully differentiated planetary body with a complex topography [2] and a multifaceted morphology including impact basins, various forms of impact craters, large troughs extending around the equatorial region, enigmatic dark material, mass wasting features and surface alteration processes [2, 5, 6]. Vesta's topography reveals extreme height differences resulting in steep slopes, locally exceeding 40° [2, 7, 8]. Those steep slopes result in craters with an unusual asymmetrical shape, where a sharp crater rim exists on the uphill side, and a subdued rim on the downhill side [7]. Impact craters on Vesta range from fresh to highly degraded, suggesting an intensive cratering history [2, 5] similar to the Moon. Vesta is believed to be the host body of HED meteorites (Howardite-Eucrite-Diogenite) (e.g., [9, 10]). The vestan surface is mainly composed of Howardite material with localized enrichments of Eucrite and Diogenite [11, 12]. The surface material consists of thick (100 meters to a few kilometers), multilayered sheets of regolith with different albedos, formed by the accumulation of ejecta from numerous impacts that have resurfaced Vesta over time [2, 13].

The dwarf planet Ceres is the largest object in the asteroid belt with a diameter of 940 km, a mean density of 2.162 ± 0.008 g/cm and a mass of 9.3835 x $10^{20}$ kg. Ceres is



IntechOpen

the only dwarf planet in the inner Solar System, which is supposed to be a relict ocean world [14, 15]. Recent observations by Dawn suggest that Ceres is a weakly differentiated body with a 40 km thick volatile-dominated crust and a rocky mantle down to a depth of 100 km comprising remnants of brines and hydrated rocks such as clays [16]. The crust is thought to be dominated by a mixture of ammoniated phyllosilicates, carbonates, salts, clathrate hydrates and no more than 30–40% water ice [17–20]. This volatile-rich outer layer is suggested to have an average thickness of 41.0 km [21–23]. The brines within the mantle of Ceres could be related to residual liquid from the freezing of a global ocean, as already proposed prior to the Dawn mission [24]. Several locations on Ceres's crust are enriched in salt compounds such as carbonates and ammonium chlorides [25]. A very large amount of the water could exist in the form of clathrate hydrates, which is conforming to geophysical conclusions for the abundance of water in Ceres's crust [17, 20, 26]. Since Ceres's globally homogenous surface is supposed to be made of material formed deep inside, a large-scale formation mechanism is suggested for that scenario. However, local heterogeneities associated with impact craters and landslides containing sodium carbonate and other salts suggest that those components are available in the shallow subsurface [26]. Sodium carbonates are found in brines of two remarkable emplacements on Ceres: Ahuna Mons [27, 28] and the bright (faculae) material in Occator crater (e.g., [19, 29]). Recent emplacement of bright deposits sourced from brines confirms that Ceres is a persistently geologically active world [19]. Generally, sodium carbonates are related to large impacts that can source deep material [26]. The most distinctive features found on both bodies are impact craters. Cratering processes on planetary bodies happen continuously and cause the formation of a large variety of impact crater morphologies. Images from the Dawn Spacecraft have revealed a diversity of impact craters, including craters with an individual appearance. The shape of an impact crater, and mainly its ejecta distribution, is the effect of a multifaceted interaction of topographic setting [30]. The majority of impact craters are more or less symmetrical and circular in shape. They display a classical circular bowl-shaped form with crater rims on the same elevation level at every azimuth and approximately parabolic interior profiles. Special topographic and subsurface conditions on both bodies have led to the development of special crater types. This chapter covers these special crater types found on Vesta and Ceres.

## 2. Asymmetric craters on Vesta and Ceres

A special type of craters occurs von Vesta. These craters show an asymmetric interior morphology and ejecta distribution and are formed on slopes. The main characteristics are a well-formed semi-circular sharp rim on the uphill side and a smooth rim on the downhill side [2, 7]. The downhill rim is covered by a local accumulation of material, whereas ejected material around the uphill rim is only sporadically distributed in thin layers (**Figure 1**). Mass wasting material is observed on the upslope inner crater walls of most craters. The majority of asymmetric craters have relatively steep inner slope angles of ~24° to 28° on the uphill side and a shallower slope angle of about 13° to 16° on the downhill side [7]. In many cases a straight line occurs on the crater floor between the oblique and the shallow side of the crater at which mass wasting material from the uphill crater wall meets the downhill crater rim material (**Figure 1**). The morphology of asymmetric craters comprises the main type described above (**Figure 1A**), as well as crater with an elongated shape in uphill direction (**Figure 1B**), v-shaped craters with one extended wall (**Figure 1C**) and craters with a lateral elongated form (**Figure 1D**) [7].

Asymmetric craters on Vesta ranging in diameter from 0.3 km to 43 km and are globally distributed. Most craters were formed on slope angles between 10° and 20°.

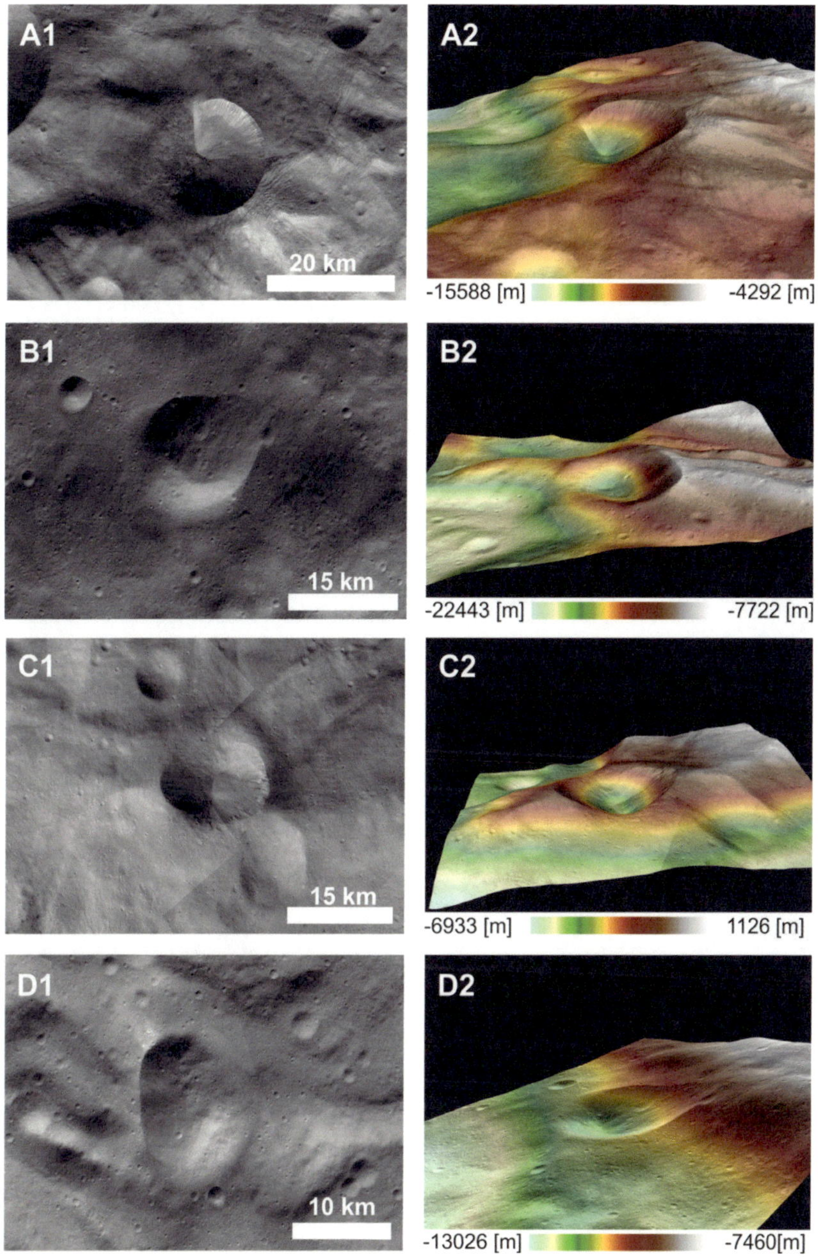


**Figure 1.**
*Examples of asymmetric craters on Vesta. A1 Antonia crater shows the classic type with a smooth downhill rim and a sharp uphill rim which is separated by a straight boundary. B1 oblique elongated crater at 6°S, 299°E. the ejecta is distributed only on the downhill rim. C1 V-shaped crater at 48°S, 129°E. D1 lateral elongated crater at 50°S, long 266°Ewith a downhill ejecta distribution at 501S, long 2661E. A2, B2, C2, D2 show the perspective view of the repective craters.*

The authors of [7] shows that the topography is the main cause for the asymmetries observed in these craters on Vesta. Numerical simulations demonstrate that the asymmetric form of these craters can be produced by an oblique impact into a slope. Additionally, the deposition of ejected material in uphill direction is prevented by the slopes, in particular by slopes >20° and results in a larger accumulation of ejecta within the crater and on the downhill crater rim. Post-impact processes are not likely because of comparable ages of crater floors and continuous ejecta [7].

Asymmetric craters are also found on Ceres. The analysis of high resolution data reveals craters similar to those on Vesta, however, with diameters from 0.30 to 4.2 km and a mean of 0.98 km the craters are much smaller [31]. The morphology of those craters shows an asymmetric crater interior with an oblique and a shallower side, as well as an asymmetric ejecta distribution. The crater reveals a semi-circular sharp and well-formed rim on the uphill side, as well as a smooth rim on the downhill side (**Figure 2**). The latter is not clearly detectable because of local accumulation of material covering the downhill crater rim [31].

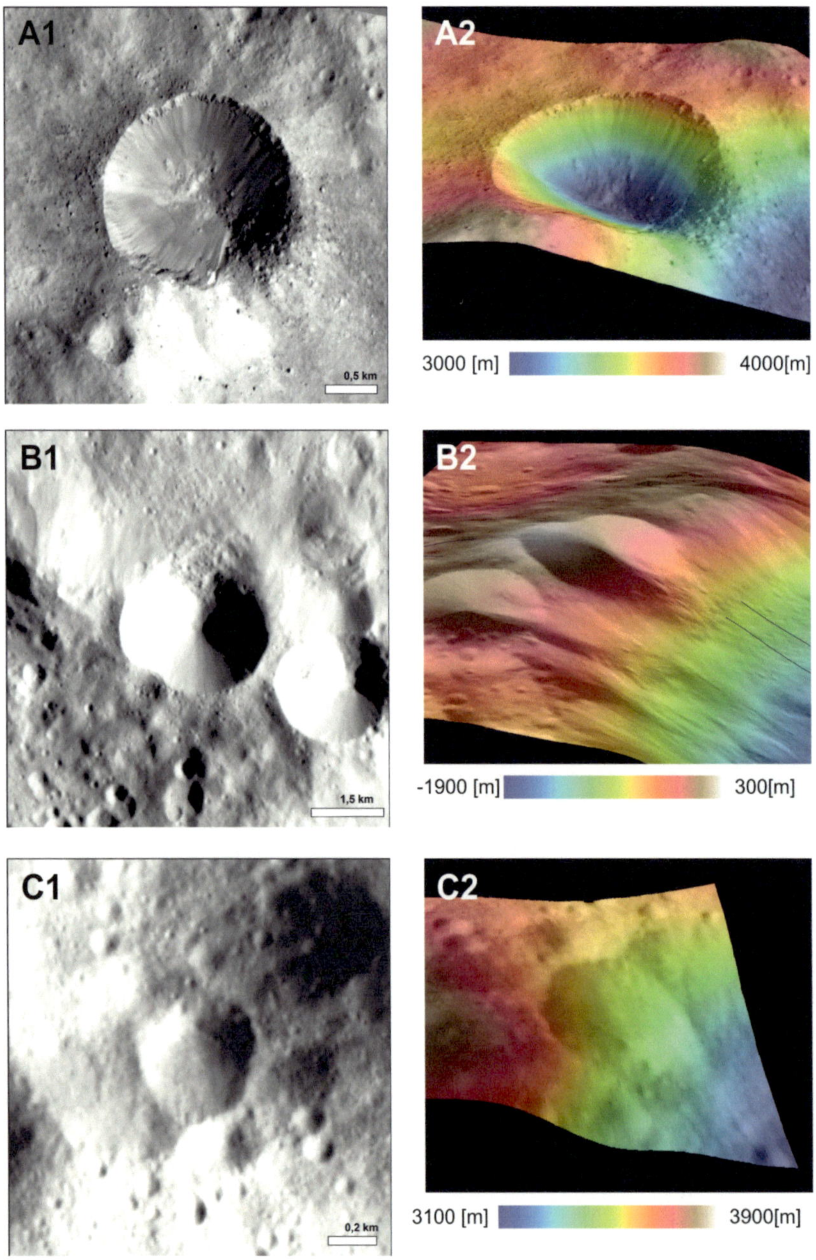


**Figure 2.**
*Examples of asymmetric craters on Ceres. A1 crater at 245.96° E and 12.57° N and B1 crater at 218.18° E and 9.55° S show a sharp uphill rim and a smooth downhill rim, covered withmass wasting material from the crater flanks. C1 crater at 245.16° E and 12.63° N shows a more elongated shape in the uphill direction than the other craters. A2, B2, C2 show the perspective view of the repective craters.*

Further similarities to Vesta's asymmetric craters comprise sporadically detected thin layers of ejecta on the uphill rim and mass wasting features on the uphill inner crater wall on most craters. Additionally, a relative straight border in the lower third through the crater, separating the oblique from the shallower crater floor [31]. Asymmetric craters are more or less circular in shape (**Figure 2A**), nevertheless there are craters showing a slightly elongated shape in uphill direction (**Figure 2C**). This crater rim seems to merge with the slope and the downhill rim is less elevated than of the typical asymmetric craters, but with the same ejecta and mass wasting material distribution. The crater floor appears wider than the others. The morphology of these craters indicating a formation on a slope crest [31]. The craters are only visible in high resolution data of the second extended mission of Dawn, and therefore, the study area is limited, although the craters are more or less homogeneous distributed over the study area, spanning around 60° N to 60° S latitude and 197° E to 265° E longitude. Most asymmetric craters are formed on slope angles between 10 and 20 degrees. [31] suggests, that the uphill and downhill material of ejecta were deposited simultaneously, and thus, not influenced by post-emplacement modifications.

Furthermore, the shape and the ejecta distribution as well as the formation on slopes are quite similar to those on Vesta. Although, the crater sizes on Ceres are much smaller for asymmetric craters, the topography is suggested to be the main cause for the asymmetries [31]. Vesta's extreme topographic differences with a total relief of ~41 km [2] have caused many craters to be formed on slopes. Moreover, the topographic differences with a total relief of ~15 on Ceres [23] is less distinctive and not as steep as on Vesta. Thus, the low variations of the topography cause lower slope angles and could have limited the crater size formation on the slopes [31].

## 3. Ring-mold craters on Ceres

Ring- mold craters (RMC) are common on lineated and lobated debris aprons, filling valleys, and concentric crater fills on Mars [32–34]. They are interpreted as impacts into ice covered by a thin layer of regolith. Ring-mold craters have diameters between 167 and 697 m and are generally surrounded by a rimless, circular moat. Furthermore, ring-mold craters show a variety of complex interior features. [33] found four morphological types of ring-mold craters: (1) a central pit or bowl; (2) a central plateau; (3) a multiring; and (4) central mound craters.

On Ceres ring-mold craters are located on the southern crater floor of Occator [35]. They predominantly appear on the lobate smooth material; a few craters are found on the terrace material as defined on the geologic maps of [36, 37]. Some ring-mold craters are located on or near the tectonic structure in the southern part of the Occator floor. Ring-mold craters show an almost circular shape seem to be subsiding into the surface, causing less elevated crater rims (**Figure 3**). Numerous ring-mold craters are degraded (**Figure 3**) and contains cracks (**Figure 3D**) or lobate material (**Figure 3E** and **F**). [35] found three classes of ring-mold craters: (1) central pit or bowl craters (**Figure 3A**); (2) central mound craters (**Figure 3B** and **C**); and (3) central plateau craters (**Figure 3B**). They show, that ring-mold craters on Ceres are comparable to those on Mars (e.g., [32, 33, 38]). Both bodies show nearly rimless craters with a circular outer moat and similar interior morphologies, like central pits or bowls, plateaus, and mounds. Moreover, ring-mold craters on both bodies are associated with flow features, lineated valley fills and lobate debris aprons on Mars and lobate materials within Occator on Ceres. The similarities of morphology and location indicate a similar formation process [35]. Although, Martian ring-mold craters with diameters between 697 (mean 225 m; [32]) and 750 m (mean 102 m; [33])

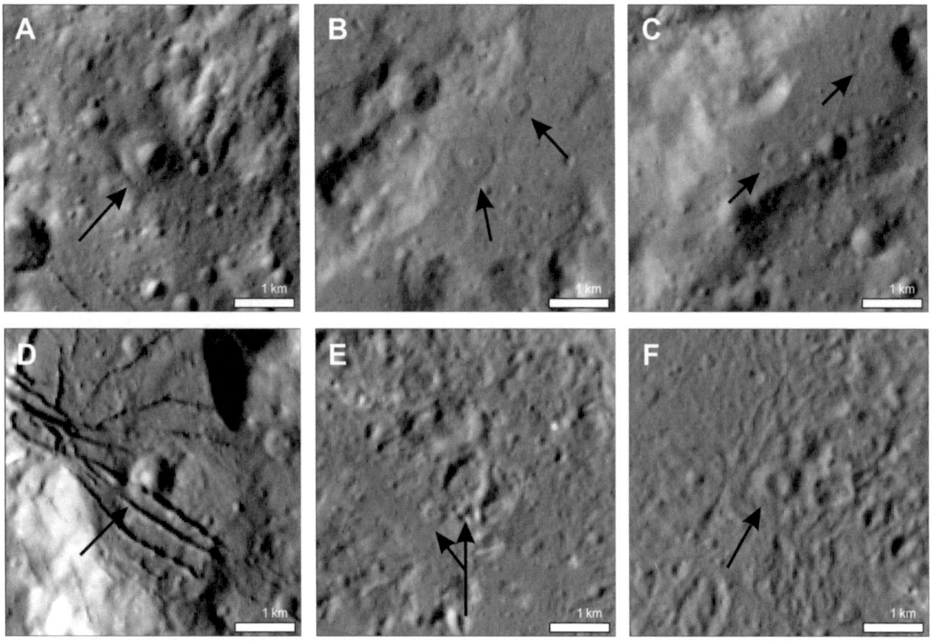


**Figure 3.**
*Ring-mold craters on Ceres. (A) shows a central pit/bowl crater. (B) shows a central plateau (bottom) and a central mound crater (top). (C) shows two central mound craters. (D) shows a ring-mold crater degraded and deformed by cracks. (E) and (F) show craters affected by flow features.*

are smaller than ring-mold craters on Ceres with diameters from ~280 to ~1,520 m with a mean of ~710 m [35]. Ponded material and lobate materials within Occator are supposed to be formed by impact melt or cryovolcanic flows (e.g., [39–41]). Furthermore, thermal modeling and gravity data suggest an extensive deep brine reservoir beneath Occator, which might have been mobilized by the heating and deep fracturing related to the Occator impact. Thus, this process would lead to a long-lived extrusion of brines and the formation of the faculae [15]. Additionally, pre-existing tectonic cracks may provide hints for deep brines migrating and dilatating the crust creating a compositional heterogeneity [15].

Due to the occurrence of frozen oceanic materials rich in sodium carbonate and ammonium chloride at several locations, [26] suggest that oceanic material is frozen in the first 10s of kilometers and possibly shallower. This hypothesis is verified by [35] using the depth to diameter ratio from [42] to estimate the minimum regolith thickness on the basis of bowl-shaped craters sizes which are adjacent to ring-mold craters. The results also conclude an overlying area thickness of several tens of meters suggesting that the smaller bowl-shaped crater do not penetrate to the ice layer. Thus, subsurface ice at Occator is at a relatively shallow depth, below a thin protective layer of regolith, and those impacts hitting the subsurface ice layer form ring-mold craters [35].

## 4. Floor-fractured craters on Ceres

Several impact craters on Ceres contains sets of fractures on their floors. Typical cerean FFCs reveal an irregular shaped rim, which is mostly deformed by slumping or sliding of the crater rim building wall terraces, and/or a central pit or peak structure [43, 44]. The morphology of most fractures is characterized by an irregular pattern with concentric and/or radial or polygonal shape, other fractures are almost straight and subparallel to parallel [44]. A common fracture structure are crevices merging into various branches or narrow fissures that conjoint into straight wide fractures. These fractures bifurcate into grand fracture groups or networks which can cover nearly half of the crater floor, this is observed at Dantu and Occator crater [44] (**Figure 4A** and **B**). A special type of floor fractures is found within Yalode crater (**Figure 4C**, **C1**). The fractures appear wider and more developed than in other craters and can be divided into two generations. They show high variation in shape, width and lengths and encompass deformational features such as en echelon structures and possible strike slip faults, dilatational jogs or tilted blocks. Such morphologic variations suggest different formation mechanism [44].

The floor-fractures are similar with floor-fractured craters (FFC) of Class 1 and 4 on the Moon (e.g., [43–45]). Depth to diameter ratios show that FFCs on Ceres are anomalously shallow similar to lunar FFCs [43]. Class 1 FFCs on Ceres includes the craters Dantu, Ezinu, Occator, Gaue, Ikapati, Azacca, Haulani, and Kupalo and shows radial and/or concentric fractures on their floors as well as central peaks or pits, similar to Class 1 FFCs on the Moon [43].

In case of Dantu and Occator, the most prominent FFCs on Ceres, an extensive set of crosscutting fractures occur the base of their southern wall (**Figure 4A2**, **B1**). The orientation of most of these fractures is concentric to the base of the crater wall, whereas the fractures at Occator are more concentrated in the southwest corner. Other fractures are orthogonal to and crosscut the concentric ones at both crater floors [43, 44]. Both craters show fractures radial and concentric to the central peak or dome/pit structure, respectively (**Figure 4A1**, **B2**). On Ceres are several linear fractures identified which are related to faculae suggesting a cryovolcanic formation [29, 44, 46, 47]. At Occator the linear fractures are related to the lobate flow fractures of Vinalia Faculae. Dantu shows more crosscutting fractures than Occator, but there are more fractures associated with the central structure [43]. Furthermore,

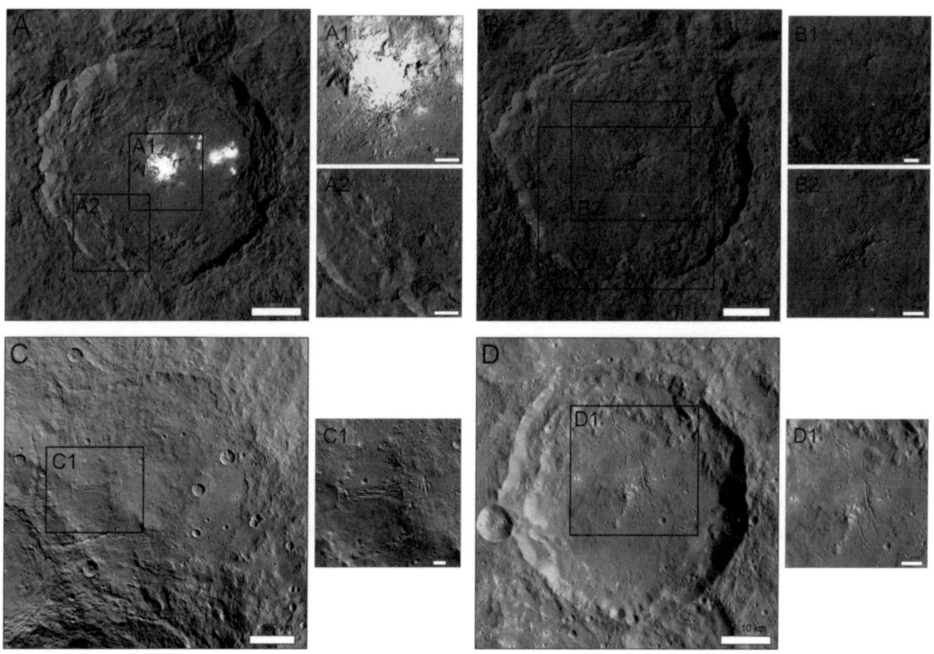


**Figure 4.**
*Examples of FFCs on Ceres. (A) Occator crater and (B) Dantu crater. A1 and B2 shows radial and concentric fractures around the central dome/pit structure. A2 and B1 shows crosscutting fractures at the base of the southern wall, respectively. (C) Yalode crater. C1 shows wider and more developed fractures which can be divided into two generations. (D) Azacca crater. D1 shows curvilinear fractures near the central peak.*

in some regions of Occator the space between fractures contains blocky fragments that seem to limit the tear faults. Other fractures are cut by slides [44]. Dantu also shows pitted fractures indicating the influence of volatile components in the subsurface [44].

Other FFCs on Ceres shows a roughly north–south trending set of curvilinear fractures and a smaller set of conjugate fractures associated with the north south fractures near the central peak in the case of Azacca (**Figure 4D**, **D1**), or contains fractures at the base of the northwestern crater wall as well as fractures orthogonal and parallel to the central crater structure in the case of Ikapati [43, 44]. Class 4 FFCs, however, are characterized by having v-shaped moats and usually hummocky floors which are more shallow than other cerean craters of their diameter. The fractures of the crater floors are less distinct. Furthermore, the craters are smaller than Class 1 FFCs [43]. Large FFCs on Ceres with diameters >50 km shows the most similarities with Class 1 lunar FFCs, while smaller FFCs are more consistent with Class 4 lunar FFCs. These results imply a similar formation of fractures due to the intrusion of a low-density material below the craters [43].

The formation of floor fractured craters (FFCs) is mainly suggested by cryomagmatic intrusion [43, 44], at which the cryomagmatic intrusion must have been trapped vertically and horizontally by weak material below the crater. Underlying reservoirs can feed such an intrusion and forming domes which uplifting and fracturing the overlying, brittle crater floor [43, 44]. Additionally, length, width and strike of the fractures vary for each crater and suggest independent formation mechanisms and imply different surface and subsurface materials. [44] also propose the following formation processes for FFCs: (1) tear-off edges in case of slumping of the crater wall, (2) cooling melting processes that lead to sinkage of the crater floor, (3) degassing, and/or (4) tectonic interactions. All four mechanism also comprises up-doming of material beneath the craters.

## 5. Conclusions

Even on small bodies impact craters reveal a multifaced morphology as caused by different formation processes. Impacts into a slope can cause the formation of asymmetric craters. Whereas steep slope angles and bigger slope sizes promote the formation of the asymmetries. However, the formation of ring-mold crater and floor-fractured crater is mainly cause by the subsurface conditions. An impact into a thin layer of regolith on top of a subsurface ice-layer lead to the formation of ring-mold crater. The occurrence of ring-mold craters can be used to detect regolith covered ice layers. Furthermore, the maximum depth of associated bowl-shaped craters can be used to estimate the depth of the ice and the thickness of overlying regolith. Floor-fractured craters is mainly proposed by cryomagmatic intrusion, which are fed by underlaying reservoirs. But also tear-off edges in case of slumping of the crater wall, cooling melting processes that lead to sinkage of the crater floor, degassing, and/or tectonic interactions are supposed to promote the formation of these crater types.

## Appendices and nomenclature

| | |
|---|---|
| FFC | Floor-fractured crater |
| HED | Howardite-Eucrite-Diogenite metroids |
| RMC | Ring-mold crater |
| Ahuna Mons | 10.48° S, 316.2° E |

| | |
|---|---|
| Antonia crater | 58.7° S, 350.78° E |
| Azacca crater | 6.66° S, 218.4° E |
| Dantu crater | 24.3° N, 138.23° E |
| Ezinu crater | 43.24° N, 195.7° E |
| Gaue crater | 30.81° N, 86.16° E |
| Haulani crater | 5.8° N, 10.77° E |
| Ikapati crater | 33.84° N, 45.6° E |
| Kupalo crater | 39.44° S, 173.2° E |
| Occator crater | 19.82° N, 239.33° E |
| Vinalia Facula | 20.2° N, 242° E |
| Yalode crater | 42.58° S, 292.48° E |

## Author details

Katrin Krohn
German Aerospace Center (DLR), Berlin, Germany

*Address all correspondence to: katrin.krohn@dlr.de

## References

[1] Russell CT, Raymond CA, Coradini A, McSween HY, Zuber MT, Nathues A, et al. Dawn at Vesta: Testing the Protoplanetary Paradigm. Science. 2012;336:684-6.

[2] Jaumann R, Williams DA, Buczkowski DL, Yingst RA, Preusker F, Hiesinger H, et al. Vesta's shape and morphology. Science. 2012;336:687-90.

[3] McCord TB, McFadden LA, Russell CT, Sotin C, Thomas PC. Ceres,Vesta, and Pallas: Protoplanets, Not Asteroids. EOS Transactions. 2006;87:105-9.

[4] Keil K. Geological History of Asteroid 4 Vesta: The "Smallest Terrestrial Planet". Asteroids III2002. p. 573.

[5] Schenk P, O'Brien DP, Marchi S, Gaskell R, Preusker F, Roatsch T, et al. The Geologically Recent Giant Impact Basins at Vesta's South Pole. Science. 2012;336:694-7.

[6] Buczkowski DL, Wyrick DY, Iyer KA, Kahn EG, Scully JEC, Nathues A, et al. Large-scale troughs on Vesta: A signature of planetary tectonics. Geophysical Research Letters. 2012;39:18205.

[7] Krohn K, Jaumann R, Elbeshausen D, Kneissl T, Schmedemann N, Wagner R, et al. Asymmetric craters on Vesta: Impact on sloping surfaces. Planetary and Space Science. 2014;103:36-56.

[8] Krohn K, Jaumann R, Otto K, Hoogenboom T, Wagner R, Buczkowski DL, et al. Mass movement on Vesta at steep scarps and crater rims. Icarus. 2014;244:120-32.

[9] McCord TB, Li J-Y, Combe J-P, McSween HY, Jaumann R, Reddy V, et al. Dark material on Vesta from the infall of carbonaceous volatile-rich material. Nature. 2012;491:83-6.

[10] McCord TB, Adams JB, Johnson TVJS. Asteroid Vesta: Spectral Reflectivity and Compositional Implications. 1970;168:1445.

[11] Prettyman TH, Mittlefehldt DW, Yamashita N, Lawrence DJ, Beck AW, Feldman WC, et al. Elemental Mapping by Dawn Reveals Exogenic H in Vesta's Regolith. Science. 2012;338:242-6.

[12] De Sanctis MC, Ammannito E, Capria MT, Tosi F, Capaccioni F, Zambon F, et al. Spectroscopic Characterization of Mineralogy and Its Diversity Across Vesta. Science. 2012;336:697-700.

[13] Jaumann R, Nass A, Otto K, Krohn K, Stephan K, McCord TB, et al. The geological nature of dark material on Vesta and implications for the subsurface structure. Icarus. 2014;240:3-19.

[14] Castillo-Rogez J. Future exploration of Ceres as an ocean world. Nature Astronomy. 2020;4:732.

[15] Raymond CA, Ermakov AI, Castillo-Rogez JC, Marchi S, Johnson BC, Hesse MA, et al. Impact-driven mobilization of deep crustal brines on dwarf planet Ceres. Nature Astronomy. 2020;4:741.

[16] Castillo-Rogez JC, Hesse MA, Formisano M, Sizemore H, Bland M, Ermakov AI, et al. Conditions for the Long-Term Preservation of a Deep Brine Reservoir in Ceres. Geophysical Research Letters. 2019;46(4):1963-72.

[17] Bland MT, Raymond CA, Schenk PM, Fu RR, Kneissl T, Pasckert JH, et al. Composition and structure of the shallow subsurface of Ceres revealed by crater morphology. Nature Geoscience. 2016;9:538-42.

[18] De Sanctis MC, Ammannito E, Raponi A, Marchi S, McCord TB,

McSween HY, et al. Ammoniated phyllosilicates with a likely outer Solar System origin on (1) Ceres. Nature. 2015;528:241-4.

[19] De Sanctis MC, Raponi A, Ammannito E, Ciarniello M, Toplis MJ, McSween HY, et al. Bright carbonate deposits as evidence of aqueous alteration on (1) Ceres. Nature. 2016;536:54-7.

[20] Fu RR, Ermakov AI, Marchi S, Castillo-Rogez JC, Raymond CA, Hager BH, et al. The interior structure of Ceres as revealed by surface topography. Earth and Planetary Science Letters. 2017;476:153-64.

[21] Ermakov AI, Fu RR, Castillo-Rogez JC, Raymond CA, Park RS, Preusker F, et al. Constraints on Ceres' Internal Structure and Evolution From Its Shape and Gravity Measured by the Dawn Spacecraft. Journal of Geophysical Research (Planets). 2017;122:2267-93.

[22] Park RS, Konopliv AS, Bills BG, Rambaux N, Castillo-Rogez JC, Raymond CA, et al. A partially differentiated interior for (1) Ceres deduced from its gravity field and shape. Nature. 2016;537:515-7.

[23] Russell CT, Raymond CA, Ammannito E, Buczkowski DL, De Sanctis MC, Hiesinger H, et al. Dawn arrives at Ceres: Exploration of a small, volatile-rich world. Science. 2016;353:1008-10.

[24] Castillo-Rogez JC, McCord TB. Ceres' evolution and present state constrained by shape data. Icarus. 2010;205:443-59.

[25] de Sanctis MC, Ammannito E, Carrozzo FG, Ciarniello M, Giardino M, Frigeri A, et al. Ceres's global and localized mineralogical composition determined by Dawn's Visible and Infrared Spectrometer (VIR). Meteoritics and Planetary Science. 2018;53:1844-65.

[26] Castillo-Rogez J, Neveu M, McSween HY, Fu RR, Toplis MJ, Prettyman T. Insights into Ceres's evolution from surface composition. Meteorit Planet Sci. 2018;53(9):1820-43.

[27] Ruesch O, Platz T, Schenk P, McFadden LA, Castillo-Rogez JC, Quick LC, et al. Cryovolcanism on Ceres. Science. 2016;353.

[28] Zambon F, Raponi A, Tosi F, De Sanctis MC, McFadden LA, Carrozzo FG, et al. Spectral analysis of Ahuna Mons from Dawn mission's visible-infrared spectrometer. Geophysical Research Letters. 2017;44:97-104.

[29] Ruesch O, Genova A, Neumann W, Quick LC, Castillo-Rogez JC, Raymond CA, et al. Slurry extrusion on Ceres from a convective mud-bearing mantle. Nature Geoscience. 2019;12:505.

[30] Elbeshausen D, Wünnemann K. The Effect of Target Topography and Impact Angle on Crater Formation — Insight from 3D Numerical Modelling. Lunar and Planetary Institute Science Conference Abstracts; March 1, 20112011. p. 1778.

[31] Krohn K, Jaumann R, Wickhusen K, Otto KA, Kersten E, Stephan K, et al. Asymmetric Craters on the Dwarf Planet Ceres—Results of Second Extended Mission Data Analysis. Geosciences. 2019;9:475.

[32] Baker DMH, Head JW, Marchant DR. Flow patterns of lobate debris aprons and lineated valley fill north of Ismeniae Fossae, Mars: Evidence for extensive mid-latitude glaciation in the Late Amazonian. Icarus. 2010;207(1):186-209.

[33] Kress AM, Head JW. Ring-mold craters in lineated valley fill and lobate

debris aprons on Mars: Evidence for subsurface glacial ice. Geophysical Research Letters. 2008;35.

[34] Kress AM, Head JW. Ring-Mold Craters on Lineated Valley Fill, Lobate Debris Aprons, and Concentric Crater Fill on Mars: Implications for Near-Surface Structure, Composition, and Age. Lunar and Planetary Science Conference; March 1, 20092009.

[35] Krohn K, Neesemann A, Jaumann R, Otto KA, Stephan K, Wagner RJ, et al. Ring-Mold Craters on Ceres: Evidence for Shallow Subsurface Water Ice Sources. Geophysical Research Letters. 2018;45(16):8121-8.

[36] Buczkowski DL, Scully JEC, Quick L, Castillo-Rogez J, Schenk PM, Park RS, et al. Tectonic analysis of fracturing associated with occator crater. Icarus. 2019;320:49-59.

[37] Scully JEC, Buczkowski DL, Raymond CA, Bowling T, Williams DA, Neesemann A, et al. Ceres' Occator crater and its faculae explored through geologic mapping. Icarus. 2019;320:7-23.

[38] Pedersen GBM, Head JW. Evidence of widespread degraded Amazonian-aged ice-rich deposits in the transition between Elysium Rise and Utopia Planitia, Mars: Guidelines for the recognition of degraded ice-rich materials. Planetary and Space Science. 2010;58:1953-70.

[39] Krohn K, Jaumann R, Stephan K, Otto KA, Schmedemann N, Wagner RJ, et al. Cryogenic flow features on Ceres: Implications for crater-related cryovolcanism. Geophysical Research Letters. 2016;43:1-10.

[40] Schenk P, Marchi S, O'Brien D, Bland M, Platz T, Hoogenboom T, et al. Impact Cratering on the Small Planets Ceres and Vesta: S-C Transitions, Central Pits, and the Origin of Bright Spots. Lunar and Planetary Science Conference; March 1, 20162016. p. 2697.

[41] Jaumann R, Preusker F, Krohn K, von der Gathen I, Stephan K, Matz K-D, et al. Topography and Geomprphology of the Interior of Occator Crater on Ceres. Lunar and Planetary Science Conference; March 1, 20172017.

[42] Hiesinger H, Marchi S, Schmedemann N, Schenk P, Pasckert JH, Neesemann A, et al. Cratering on Ceres: Implications for its crust and evolution. Science. 2016;353(6303).

[43] Buczkowski DL, Sizemore HG, Bland MT, Scully JEC, Quick LC, Hughson KHG, et al. Floor-Fractured Craters on Ceres and Implications for Interior Processes. Journal of Geophysical Research (Planets). 2018;123:3188-204.

[44] Krohn K, von der Gathen I, Buczkowski DL, Jaumann R, Wickhusen K, Schulzeck F, et al. Fracture geometry and statistics of Ceres' floor fractures. Planetary and Space Science. 2020;187:104955.

[45] Schultz PH. Floor-fractured lunar craters. Moon. 1976;15:241-73.

[46] Buczkowski DL, Williams DA, Scully JEC, Mest SC, Crown DA, Schenk PM, et al. The geology of the occator quadrangle of dwarf planet Ceres: Floor-fractured craters and other geomorphic evidence of cryomagmatism. Icarus. 2017.

[47] Quick LC, Buczkowski DL, Scully JEC, Ruesch O, Castillo-Rogez J, Raymond CA, et al. Thermal and Compositional Evolution of a Brine Reservoir Beneath Ceres' Occator Crater: Implications for Cryovolcanism at the Surface. Lunar and Planetary Science Conference; March 1, 20182018.